New Physical Phenomena Responsible for Ball Lightning

V. P. TORCHIGIN

CONTENTS

Introduction

We are going to present a set of new physical phenomena that we have managed to disclose without leaving a computer. We took into account the thought of the Russian Nobel prize winner Peter Kapitsa that Ball Lighting (BL) is a small window in the great world of new physical phenomena. It is not a simple task to find the window because scientists of whole world cannot account for two centuries the nature of Ball Lightning. Above 200 various theories are proposed, above 2000 papers and reports are published but simple questions cannot be answered. Why may BL move upwind, how does BL penetrate in a room through windowpanes, chimneys, small splits and holes, why does BL radiate white light, the spectrum of which corresponds to the temperature of several thousand degrees and is relatively cold at the same time? How does BL catch up a flying airplane and penetrate within its salon? How can BL store extremely great amount of the energy? The list of the questions may be easily continued. We need to agree with the conclusion of Sagan - authors of the last book about Ball Lightning [Sagan 2004] entitled "Ball Lightning: Paradox of Physics." All theories have one thing in common - none work.

There is an understanding that a majority of the theories ought to be withdrawn and efforts ought to be concentrated on hypothesizes which satisfy certain requirements.

The first requirement is a BL ability to penetrate through windowpanes. As is known, any particles such as molecules, electrons, ions, clusters, etc. cannot penetrate through glass. Because of this, either BL does not contain such particles or BL can generate such particles at opposite sides of a windowpane.

The second requirement is BL ability to move in the direction that does not coincide with the direction of a wind that is BL ability to be not blown down by air streams. Any object consisting of material particles is blown down by air streams. On the contrary, BL can accompany a flying airplane. In this case, BL is not blown by the wind, the speed of which surpasses the speed of the greatest hurricane. At present, nobody can imagine the object that satisfies these requirements. It is necessary at first to invent such object before investigating its properties.

Existence of the self-confined light in nature

In 2002, we put forward a hypothesis [Torchigin 2002] where BL is considered as a self-confined light. This is perfect unusual at first glance object in a form of a spherical self-confined light bubble or ball light. Like any bubble, Ball Light has a shell. Unlike a conventional soap bubble and the ball light considered in [Ignatovich 1992], no excess pressure is present within the Ball Light volume. Ball Light is the shell itself. Ball Light shell is a compressed air where an intense light circulates in all possible directions. The refractive index n of the compressed air is greater than that of the surrounding air. In fact, the thin film of the compressed air is a thin-film-planar-lightguide which curvature is different from zero. In turn, the intense light produces the electrostriction pressure in any optical medium, in particular, in the air where the light propagates. The electrostriction pressure is proportional to the light intensity and tends to near the air molecules close together. Thus, the Ball Light shell is a system of the compressed air and intense light. The compressed air provides a confinement of the intense light and the intense light provides a confinement of the compressed air. In other words, Ball Light shell is a self-confined light in the nonlinear optical medium in a form of conventional air. The air pressures within the volume of Ball Light and outside the Ball Light shell are the same and are equal to the normal atmospheric pressure. Planar light guides like thin Ball Light shell are a basis of up-to-date integrated optics [Kogelnik 1988]. They confine a light propagating within them from radiation in free space.

Common sense dictates that the self-confined light does not exist. No one ever mentioned this stupidity. But a particular case of the self-confined light is known since 1971. This is an optical space soliton in a form of a plane light beam of width w that is propagating in a nonlinear medium, the refractive index of which increases with an increase of the light intensity. In this case, the phenomenon of self-focusing takes place and the width of the beam decrease gradually. On the other hand, the width of the beam increases due to the phenomenon of diffraction divergence. These two phenomena compensates each other and the width of the beam is not changed. We can say that the light beam confines its width or, in other words, is self-confined in one direction.

The ball light can be considered as a generalization of flat incoherent optical spatial soliton propagating in the optical nonlinear medium that is the usual atmospheric air. The ball light can be obtained by transformation of such soliton in 2 stages. In fig.1a, a fragment of a conventional plane incoherent soliton is shown. At the first stage, it is possible to replace a plane beam of intensity I propagating in one direction by a set of the plane beams propagating in all possible directions and having the same total intensity I (fig.1b). Such replacement is correct because the total intensity of light remains unchanged, as the intensity of a superposition of incoherent beams is equal to the total intensity of these beams. In this case, an influence of light on the nonlinear optical medium in fig.1a is the same, as in fig.1b.

At the second stage, the curvature of the incoherent spatial soliton in fig.1b increases from zero up to some finite size R^{-1} (fig.1c). Thus, we obtain a spherical spatial soliton. Unlike an infinite spatial soliton a in fig.1a, the spherical soliton borrows enclosed area of space, and its diameter is equal $2R$. No one suspected about an existence of a spherical spatial soliton in nature until now or, in our terminology, about an existence of ball lights.

On assumption that the BL is the experimental confirmation of the existence of such solitons, our task to study properties of Ball

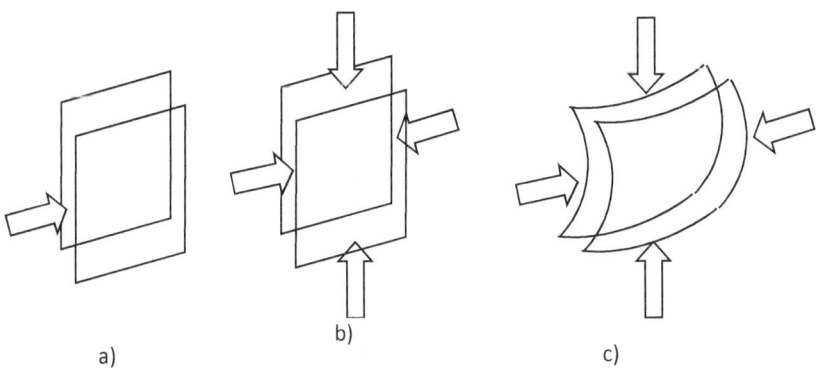

a)   b)   c)

Fig.1. Steps of transformation of a fragment of plane soliton into spherical one.

Light becomes much easier. We are in a situation where the answer is known in advance, and we have to look for a solution only.

Propagation of waves along the closed trajectory resembles the propagation of waves of a so-called "whispering gallery" mode (WGM) along a closed waveguide. The waves of the WGM type are known largely as resonance modes used in optical resonators based on glass microspheres. The WGM wave can be represented as a traveling wave moving along the microsphere equator and reflecting repeatedly from the spherical glass-air boundary. Such waves can be excited even in barrel-shaped segments of an optical fiber as well as in cylindrical waveguides. In the latter case, they are termed the tunneling modes [Govar 1989].

Certainly the main argument of existence of the self-confined light in a form of Ball Light is not out theoretical considerations, but the existence of Ball Light in nature in a form of natural Ball Lightning. Evidence of this is a striking coincidence of behaviors of Ball Lightning derived from evidences of eyewitnesses and Ball Light derived from known physical laws.

The following puzzles of the natural Ball Lightning behavior have been explained: ball lightning behavior near the earth's surface, the reasons of its movement in a horizontal direction at a small distance from the earth's surface [Torchigin 2003], possibility of ball lightning motion upwind, BL property to accompany airplanes and penetrate in their salons [Torchigin 2004 Phys. Lett.], ball lightning penetration in rooms through splits in the walls and through window panes [Torchigin 2003].

Thus, the theory of Ball Light can be considered as a window in the world of new unknown phenomena. Certainly, the most main phenomenon is an existence in the nature Ball Light. Besides, we have disclosed several other unknown phenomena.

This is the new type of optical nonlinearity in the mixture of gases that surpasses the known Kerr-like nonlinearity and the nonlinearity due to the electrostriction pressure.

This is the self-organization of the intense light in gases. This phenomenon has been discovered at the analysis of the appearance of miniature luminous ball in the vapor of the liquid nitrogen that is illuminated by the intense light.

This is the new phenomenon where the circulating light is produced gradually due to the exothermal chemical reaction in bouncing balls. Below we present these unknown phenomena that were derived from our notions about optical nonlinear effects in gases.

New mechanism of optical nonlinearity in gaseous mixture

Spectra of the AOs generated by means of gas discharges caused by powerful radio-frequency radiation, were investigated in work [Barry 1980]. The following features of AO spectra have been noted. The AO spectrum corresponds to the spectrum of impurity, rather than the spectrum of the gas in which AO is formed. The AO spectrum corresponds basically to a spectrum of carbonic gas, and also contains some lines of metals of which electrodes have been made. AO color specified also that there are some traces of the dioxide of nitrogen NO_2. However, it was impossible to analyze spectrum NO_2 on a background of spectrum CO_2. We should notice, that a refractive index of carbonic gas n_{CO2} = 1.0004197, and dioxides of nitrogen n_{NO2} = 1,000515. At the same time refractive indexes of air, nitrogen and oxygen in which AO were formed, are in an interval from 1.00025 till 1.00028. Thus, the AO spectrum consists of those components of a gas mixture, which refractive indexes are the greatest.

The increase in a refractive index in Ball Light shell can occur not only due to compression of air, but also due to drawing in the Ball Light shell gas components with the maximal refractive indexes. The light energy for increasing the refractive index at some value Δn is required smaller in the latter case than in the first one. Thus, studying of abnormal AO spectra has allowed to disclose a new type of optical nonlinearity in gases.

From the presented consideration follows, that natural BL also can have a shell consisting of various vapors. Really, there are numerous evidences that such BLs arise often enough. Occurrences BL are connected often by eyewitnesses with distinctly felted smells. More often, these smells were defined as a smell of sulfur and ozone. In several cases, the smell was compared with a smell of the dioxide of nitrogen. All these gases have the refractive index, which Δn was approximately twice greater than that of usual air.

Some observations indicate that an initial linear lightning can generate BL, having struck in any firm substance and having evaporated it. An occurrence of a fiery sphere that slides from a tree prone to a lightning strike is typical enough [Stakhanov 1979]. The fact of detection of pitch in residues from BL appeared at a strike of a linear lightning in a tree is in a good agreement with presented picture

In other evidence, it is reported about fireball on the earth, which exploded after 30 seconds. The ball appeared in an extremely strong thunderstorm. There were tar residues after the explosion. They have the sulfur smell and were so hot that they burned eyewitness hand through 10 minutes after the explosion. From hitting at to this substance his finger became yellow.

There are numerous evidence that BL can contain material objects. For example, Stakhanov describes case where the hot black mass is observed after the disappearance of BL that arises from the linear lightning that strikes into a tree.

Consider now a condition of thermodynamic balance of a gas mix in which the intensive light propagates. For simplicity, we will consider a mix consisting of two components only which refractive indexes are equal n_a and n_b, respectively, and $n_b > n_a$. Let their initial relative concentration be equal, respectively, to z_a and z_b ($z_a + z_b = 1$). As is known, a certain density energy W_m for separation of components of a gas mix is required. This density energy is determined by the following expression:

$$W_L = T \Delta S \tag{1}$$

where T is the temperatures of the mixture, ΔS is an increase of the entropy of gas mixture [Naschokin 1969]. But ΔS is equal to zero at adiabatic process where no heat is added to the system. We consider namely such situation. This means that a change of the energy of the gas mixture at conditions closed to adiabatic ones is closed to zero. In this case, a small light density can force a full separation of the gas mixture. On the other hand, at the full separation where only molecules with n_b are located in the region where the intense light propagates, an increase of the refractive index is the following

$$\Delta n = n_b - [z_b n_b + (1 - z_b n_b)] = (1 - z_b)(n_b - n_a)$$ (2)

Usually in experiments $z_b \ll 1$ и $(n_b-n_a)>(n_0-1)$. In this case $\Delta n > n_0-1$. This means that at a relatively small intensity we can obtain a change Δn which is greater than that at the electrostriction pressure $P_L > P_0$. The electrostriction pressure P_L is given by

$$P_L = \tau \frac{d\varepsilon}{d\tau} W_L$$ (3)

Where τ and ε are the air density and permittivity, respectively. The air permittivity $\varepsilon = 1 + \Delta\varepsilon$ where $\Delta\varepsilon \ll 1$. In this case we have $\Delta\varepsilon/\tau = const$ and, therefore $d\varepsilon/d\tau = \Delta\varepsilon/\tau$. At normal conditions the air pressure is equal $P_0 = 10^5$ Pa and $\Delta\varepsilon = 5.4 \ 10^{-4}$. The electrostriction pressure equaled to P_0 is achieved at the density of light energy $W_L = 1.8 \ 10^8$ J/m^3. This corresponds to the light intensity $I_e = W_L \ c = 5.4 \ 10^{12}$ W/cm^2. The light intensity I_s required to separate the gas mixture and to increase the refractive index in accordance with (1) depends on a degree of "adiabaticness" of separation process. As follows from experiments, the light intensity in the gas discharge is sufficient for separation. In this case I_s is smaller than I_l by several orders of magnitude and, therefore, the degree of nonlinearity of the gas mixture may be extremely great.

Consider now inertial properties of nonlinearitics in gases. As is known, the time constant of Kerr nonlinearity is about 10^{-12} s. The time constant of the electrostriction nonlinearity is equal approximately to the time that is required for a sound wave to pass the distance equal to the light beam radius. At the radius of 10 μm and sound speed of 340 m/s we obtain for the time constant $\tau \approx 30$ ns. This is comparable or greater than the width of power laser pulse. Transient processes for the considered type of nonlinearity are connected with a suction of molecules in the region where intense light exists that is with the phenomenon of transport. In this case the time constant increases by orders of magnitude because time of transient processes is determined by the time when a steady state is set up in the whole volume of gas mixture but not only in the region where intense light exists. Thus, the picture considered above is true for the light of significant duration.

Usually the light power is extremely great and is measured in Megawatts. Such light may be in a form of pulse only. The light power at discharge of capacitor battery with the stored energy about 1 KJ in time about 5 ms is equal to 0.2 MW. In this case, the light propagates in all possible directions and seemingly, there is no sense to speak about separation. Nevertheless, as will be shown, the separation takes place in this case too. It turns out that an intense light propagating in gas mixture is instable.

Self-organization of the intensive light in gases

In optical mediums, in particular, gases, several types of optical quadratic nonlinearity are known. They manifest themselves in quadratic dependence of the refractive index of the medium on amplitude of the light wave propagating in the medium. A degree of nonlinearity is characterized by the factor n_n in expression

$$\Delta n = n_2 I, \tag{4}$$

where n_2 determines an increase in refractive index Δn of a medium at propagation the light of intensity I within it. The most known are Kerr nonlinearity at which the increase in n is connected with orientation of molecules in a field of a light wave, and electrostriction one which is connected with compression of the medium, within which the light is propagating. The compression of the optical medium entails increases in its density and its n.

It appears that in mixes of gases, which components have various n, there is one more mechanism of occurrence of optical nonlinearity at which molecules of gas with the greatest n are involved in area of a light beam. As a result, the refractive index n of the mix in this area increases. Thus, n within a light beam depends on its intensity. As the energy connected with separation of molecules in a gas mix to increase in its n can be smaller than the energy connected with compression of the gas to increase n in the same degree, the index n_2 in the first case is greater than that in the second one.

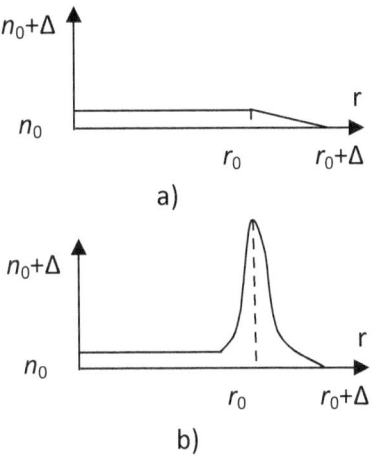

Fig. 2
Dependenceoftherefractiveindex$n$within ballonthedistance$r$fromitscenter. (a) initial state; (b) after self-organization

We shall assume that simultaneous action of all these mechanisms of nonlinearity is taken into account in value of the index of nonlinearity n_2 in Eq.(4).

Consider a fluctuation of gas density appeared as a result of a chaotic motion of gas molecules. Such fluctuations appear in any gas constantly and they are responsible for the known optical effect of molecular light scattering. Let for the sake of simplicity a form of the fluctuation is a ball which radius r_0 is essentially greater than light wavelengths are. Since the refractive index n of a gas is proportional to the gas density, the dependence of n on the distance r from the ball center can be presented by the curve shown in fig.2a. It is supposed that the spherical layer of thickness Δr is a transitive layer where n decreases linearly from $n_0+\Delta n$ within the ball to n_0 outside it. Let the intense light radiated by excited atoms is homogeneous in the region where the fluctuation is located. This means that there are no preferable directions and the densities of the light energy in all points are identical. These assumptions simplify considerations significantly because the light intensity does not depend on coordinates.

Trajectories of light beams propagating in a cross-section of the ball by plane z=0 are shown in fig.3. Similar picture takes place in any cross-section because of central symmetry of the considered configuration. As is known any light beam propagating in an inhomogeneous optical medium bends in the direction where the refractive index increases. The longer the light beam propagates in an inhomogeneous medium the greater its bending is. In fig.3, this

13

concerns the beams which propagate in the transitive layer of thickness $\Delta=w$ perpendicular to the ball radius. We will consider such beams only.

The radius R of curvature of the beam is determined as follows

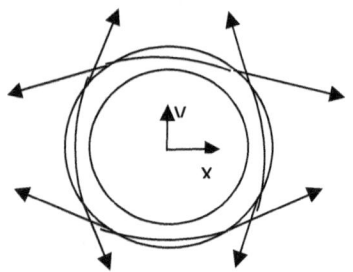

$$R^{-1}= \operatorname{grad}(n) \cos\theta \qquad (5)$$

where θ is an angle between directions of $\operatorname{grad}(n)$ and the beam. In the considered case

$$R^{-1}\approx\Delta n/\Delta r$$

Fig.3. Trajectories of light beams in the transitive layer

$$(6)$$

Clearly, that the bend of beams to the center of the ball entails an increase in the light intensity inside the transitive layer. Indeed, in this case any cross-section of the transitive layer is crossed by the greater number of light beams entered the transitive layer as compared with the case where $R^{-1}=0$. One can see that an increase in the light intensity is proportional to the length of the trajectories of beams that are located in the transitive layer.

An increase in the light intensity in the transitive layer can

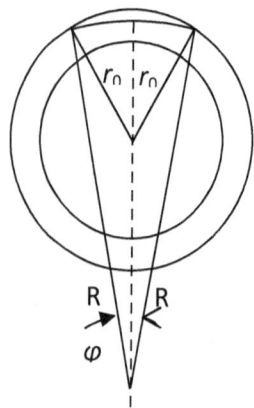

cause an increase in its refractive index n if the optical medium is nonlinear. In turn, the increase in n entails an increase in the light intensity and so on. As a result, a ball light can appear where a character of dependence $\Delta n(r)$ is shown in fig.2b.

There are new nonlinear optical effects discovered at analysis of experimental results obtained at investigations of AOs properties. First of them takes place in gas mixtures and is connected with the fact that molecules of gas mixture component with maximal n

Fig. 4. Schematic diagram for calculation of length of a light beam in the transitive layer

are involved in the area where an intense light propagates. The second one is connected with the fact that electrons of low temperature plasma are pushed out from the area where an intense light propagates. As a result, n in the area increases. The electrons attract positive ions and, therefore, plasma is pushed out from the area.

We will suppose that an action of all these mechanisms results in an increase in n by δn at increase of the light intensity by ΔI and relation (1) takes place

Let us determine an increase in the light intensity in the transitive layer. We have from the drawing shown in fig.4,

$$Cos\varphi = 1 - \Delta r(\Delta r/R)/(R-r_0+\Delta r) \tag{7}$$

Under assumption that $\varphi < 1$ and $Cos\varphi \approx 1 - \varphi^2/2$ we have from (7) and (6)

$$\varphi \approx 2^{1/2}(\Delta n) (1-\Delta n(r_0-\Delta r)/\Delta r)^{-1/2} \tag{8}$$

Taking into account that the index C of increase of the light intensity in the transitive layer is proportional to φ, parameter C can be expressed as follows

$$C = [(\Delta n + \delta n)/\Delta n][1-(r_0-\Delta r)\Delta n/\Delta r]^{1/2} \times$$
$$[1-(r_0-\Delta r)(\Delta n+\delta n)/\Delta r]^{-1/2} \tag{9}$$

As is seen, $C=1$ at $\delta n=0$ and C tends to infinite as the term in the last square brackets tends to zero. This takes place if the term $(\Delta n+\delta n)/\Delta r$ tends to $(r_0-\Delta r)^{-1}$. Taking into account (6), we can conclude that C tends to infinity when the radius of curvature $R=(\Delta n+\delta n)/\Delta r$ for light beams in the transitive layer tends to the radius of the transitive layer r_0

Denoting the light intensity in the transitive layer by I_T, we have $I_T=CI_0$ where I_0 is the light intensity in surrounding space. Then from (9) we obtain

$$I_T=I_0[(1+n_2I_T/\Delta n] [1-(r_0-\Delta r)\Delta n/\Delta r]^{1/2}$$

15

$$[1-(r_0-\Delta r)(\Delta n+n_2 I_T)/\Delta r]^{-1/2} \qquad (10)$$

This expression determines dependence between I_0 and I_T. For the sake of illustration, a particular case of the dependence is presented in fig.5. An increase in I_T occurs more quickly than increase in I_0. This is explained by the fact that the increase in I_0 is accompanied by an increase in the length of the light trajectory resided in the transitive layer. When the length becomes equal to the length of the transitive layer, the light beam trajectory becomes closed and the light beam can circulate repeatedly around the transitive layer. The length of the beam trajectory resided within the transitive layer increases many times over in this case. The transitive layer shows itself as a planar lightguide which curvature is different from zero. As is known, similar lightguides can confine the light launched within them. The light can circulate in the lightguide during some time after termination of gas discharge when the intensity I_0 becomes equal to zero. Such situation takes place for autonomous objects produced in 19 century by Plante, Toepler, Ledus and others [Barry 1980].

A wave approach is more appropriate for analysis of light circulation than the beam approach applied above. In this case ought to consider light waves of whispering gallery type circulating in the transitive layer in all possible directions. Such waves are used in optical resonators of whispering gallery type [Spillane 2002]. Glass balls of several ten micrometers in diameter are used as such resonator. The refractive index of glass is about $n\approx1.45$. In this case, the difference between the refractive indexes of glass and surrounding space $\Delta n\approx0.45$. This is very great difference that provides a safe confinement of light waves within glass balls. The difference Δn in a Ball Light ball is essentially smaller than that in a glass ball. The more the difference the better

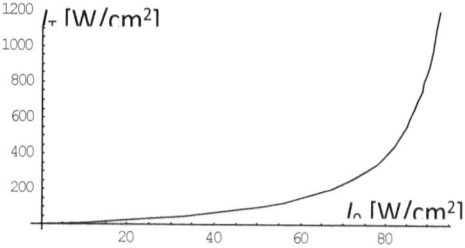

Fig. 5.Dependence of the light intensity within transitive layer I_T on the surrounding intensity I_0 ($r0=10^{-5}$m, $\Delta r=10^{-6}$ m, $\Delta n=10^{-4}$, $n_2=10^{-5}$ cm^2/W)

16

confinement is. Since the difference is proportional to the light intensity in the transitive layer, the more light intensity the more confinement is. This conclusion corresponds to the conclusion derived from the analysis based on the beam approach. Ought to add a single specification that a confinement increases continuously with an increase in background light intensity I_0 and there is no jump at the moment when a light beam in the transitive layer becomes closed one.

Since AOs disappear immediately after ceasing gas discharge, the light intensity I_T in the transitive layer is insufficient to form a lightguide with small radiation losses. Nevertheless, the self-organization of intense light takes place in Ball Light balls also because occasional fluctuation of gas density is transformed in the ball light. The light intensity I_T in the shell of the ball light can be essentially greater than the background light intensity I_0.

At $I_0>I_{00}$ the situation is perfectly different. Trajectories of light beams become closed ones. A process of accumulation of light within the transitive layer considered above takes place. After a disappearance of the homogeneous light, the light within ball light continues to circulate and is scattering gradually. Ought to underline that a propagation of the light within the transitive layer is connected with nonlinear phenomena. The propagation differs essentially from the propagation of light in a linear medium with the same distribution of the refractive index n in space [Torchigin 2005].

There are two sources of the light for accumulation. First, it is excited atoms within the transitive layer. Light beams radiated perpendicularly to the radius begin to circulate in the layer and by doing so they increase the light intensity in the layer. Second, it is the background light radiation that surrounds the ball light. The background light beams that are tangent to the ball light surface are drawn in the ball light. The main problem is providing prolonged accumulation of light in the bubble. Since the temperature in the region of gas discharge is greater than that in surrounding space, the air density in the region becomes smaller than that in the surrounding space. Since a ball light moves in the direction of the gradient of the air density, the ball light leaves the region and the light accumulation ceases. Experimentalists discovered several ways to overcome these problems. Their means are analyzed in the last section.

As is known, such optical waveguides also can guide and limit light entered into them. Such light can circulate in an optical waveguide during some time after the termination of the gas discharge when intensity of background light I_0 becomes equal to zero. This situation takes place for AOs with noticeable lifetime. Consideration of circulation of light in an optical waveguide is expedient on the basis of wave approach rather that beam approach used before. In this case, it is necessary to consider waves of type of the whispering gallery, circulating in every possible directions in a transitive layer. Such waves are widely used in optical resonators [Spillane 2002].

As such resonator the glass ball in diameter in some tens micrometers is used. The refractive index of glass is approximately equal $n=1.45$. In this case a difference in refractive indexes of a medium and surrounding space $\Delta n=0.45$. It is a very great difference that provides reliable confinement of light waves in the shell. Earlier it has been shown, $\Delta n \approx 0/01$ is sufficient for reliable confinement of light in a ball of radius of several centimeters and the greater the difference is the better confinement. This conclusion coincides with a conclusion received above on the basis of the beam approach.

It is necessary to add one refinement. From consideration on the basis of the wave approach follows, that confinement of light improves continuously at increase in the intensity of background radiation and any jumps is not observed during this process when the trajectory of a light beam in a transitive layer becomes closed. If ball lights disappear right after the terminations of the gas discharge, the intensity of light I_T in the transitive layer is insufficiently high to provide small radiating losses. Nevertheless self-organizing of intensive light in this case takes place also, and casual fluctuations of density of air can be transformed to ball lights. The light intensity I_T in the transitive layer can essentially surpass the background intensity I_0.

These theoretical considerations are confirmed by the mentioned experiments. An occurrence of AO near the surface of liquid nitrogen and penetration of AO in liquid nitrogen can be explained as follows. The boiling point of liquid nitrogen is equal to -195 C^0 and a boiling point of oxygen is equal to -182 C^0. Therefore, oxygen in the atmosphere at a certain distance above the liquid nitrogen is in the critical opalescence.

Miniature ball lights with a shell of liquid oxygen arise under the action of intense light due to the self-organization of intense light. Then they move along the gradient of the refractive index of air, in which they are located. The temperature gradually decreases with approaching the surface of liquid nitrogen. Therefore, there is a significant gradient of the air refractive index directed toward the surface of liquid nitrogen, and AO is moving toward the surface. Nearer to the surface, AO starts to evaporate the liquid nitrogen in the same way as it evaporates the metal foil when it is approached thereto. The temperature of the gas in the layer between AO and the liquid nitrogen is close to the temperature of liquid nitrogen, and the refractive index n of the layer significantly exceeds the refractive index of the surrounding space. As a result, the seeking area with a maximum refractive index, AO goes to the coldest region and "burns" the hole in liquid nitrogen. The depth of the hole gradually increases. Ultimately, AO penetrates fully into the liquid nitrogen and formed around itself a coat of nitrogen gas. Apparently, the luminescence of the volume of liquid nitrogen just explained by the glow of plurality of AO of small diameter penetrated into liquid nitrogen.

With reducing the diameter of the AO, when AO diameter becomes comparable with the light wavelength, the radiation losses in the AO shell consisting of liquid nitrogen begins to grow significantly. In this case, the radiation loss for the light in the long wavelength region of the spectrum are significantly higher than in the short one [Oraevskiy 2002]. Therefore, the lifetime of the AO of a small diameter is relatively small (5 s) as compared to the lifetime of the AO of large diameter (30 s). Furthermore, this explains the fact that the light in the process of AO life gradually changes its color from bright white to deep purple. The long-wavelength part of the spectrum is radiated by pretty quickly. Thereafter AO radiates over a relatively long period of time a blue-green color. As this radiation in AO is only part of the short-wave radiation, the color turns purple. The fact that the radiation gradually fades indicates that AO shell is liquid. If the shell is composed of highly compressed air, AO would cease to exist suddenly with a characteristic bang. This phenomenon is explained in [Torchigin 2007].

Miniature ball lights with a shell of liquid oxygen arise under the action of intense light due to the self-organization of intense light. Then they move along the gradient of the refractive index of air, in which they are located. The temperature gradually decreases as approaching the surface of liquid nitrogen. Therefore, there is a significant gradient of the air refractive index directed towards the surface of the liquid nitrogen and AO is moving towards the surface. Approaching to the surface, AO starts to evaporate the liquid nitrogen in the same away as it is burned through the metal foil. The temperature of the gas in the layer between AO and the liquid nitrogen is close to the temperature of the liquid nitrogen, and the refractive index n of the layer significantly exceeds the refractive index of the surrounding space. As a result, seeking the region with a maximum refractive index, AO goes to the coldest region and "burns" the hole in the liquid nitrogen. The depth of the hole gradually increases. Ultimately, AO penetrates fully into the liquid nitrogen and the coat of a nitrogen gas is formed around AO.

Apparently, the luminescence of the volume of liquid nitrogen is just explained by the glow of the plurality of AO of small diameter penetrated into the liquid nitrogen.

Capture of external light by self-confined light

The density of the light energy W_L within a gas discharge may be estimated as follows. In accordance with the Stefan-Boltzmann law the total volume density of radiation is the following $W_L=aT^4$, where $a=7.56 \ 10^{-16}$ J m^{-3} K^{-4}. At $T=3 \ 10^4$ K^0 we have $W_L= 612$ J/m^3. It is essentially smaller than that required to increase $\Delta n=n_0-1$ by several times to provide a confinement of the light within Ball Light, where $n_0=1.000277$ is the refractive index of the air at normal conditions. There should be some other mechanisms that provide an increase in the density of the light energy. One of such mechanism has been considered in [Torchigin 2007] where was shown that about 10% of the light energy E radiated by an excited atom located within Ball Light shell is preserved in the shell. Thus, the light energy stored within Ball Light increases linearly by $0.1E$ for each period of rotation $\tau \approx 10^{-10}$ s. Since an accumulation of the energy at the erosive gas discharge lasts several milliseconds, the light energy within Ball Light can be increased by a factor of 6 orders of magnitude. This conclusion is confirmed by optimal conditions for production of Ball Lights. Optimal duration must be as long as possible. Increase in the temperature within the discharge space entails an increase in the air pressure. As a result, molecules of the air leave the space and the air density decreases. Moving in the direction where the air density is maximal, Ball Lights leave the space too and process of accumulation of the light energy within Ball Lights ceases. Additional measures ought to be undertaken to retain Ball Light in the discharge space. For example, experimentalists have discovered that so called erosive gas discharge is favorable for AO production. In this case, evaporation of electrodes or walls of a discharge camera provides a delivery of new portions of gas in the discharge space and an increase in n in such a way.

It turns out that there is once more mechanism that provides an accumulation of the light energy within Ball Light. An accumulation is connected with properties of a space soliton to draw in light beams propagating parallel to the soliton at some small distance from it. The same mechanism provides a conversion of a plain light beam propagating in a Kerr-like nonlinear optical medium in a stable space soliton [Haus 1984].

Consider initially this property in the case of known classical plane space soliton. It has been shown that a light beam which sizes of the cross-section are equal to w along the x-axis and infinity along the y-axis propagating along the z-axis acquires a stable profile which intensity is described by the square of function [Haus 1984]

$$u(x,z) = \sqrt{\frac{n_0}{k_0^2 n_2}} \eta \frac{Exp(-j\eta^2 \frac{z}{2k_0})}{Ch(\eta x)} e^{-jk_0 z}$$

(11)

It is supposed that the light wave is propagating along the z axis, is infinite along the y axis, and the profile along the x axis is given by Eq. (11). The width of the profile is equal to $w=1/\eta$ and is determined from the condition

$$J(w)=2\pi \tag{12}$$

where

$$J = \sqrt{\kappa/2} \int_{-\infty}^{\infty} u_0 Ch^{-1}(x/w)dx \tag{13}$$

$\kappa=2k_0^2 n_2/n_0$, $k_0=2\pi/\lambda_0$, λ_0 – is the light wavelength in vacuum, n_0 is the refractive index of the medium for a light wave of small amplitude, n_2 is the index of nonlinearity determined by the relation $\Delta n=n_2 I$, Δn is an increase in n under action of the light wave of I intensity, $u_0^2=I_0$, I_0 is the maximal intensity at $x=0$.

One can see from (12), (13) that

$$J\sim I_0^{1/2}w \tag{14}$$

and, therefore, the greater I_0 the smaller the width w of the soliton. There is a very important property of beams from which soliton is formed [Hays 1984]. If J in (13) for any light beam at some $z=z_0$ is in the range $\pi<J<3\pi$, the parameters w and I_0 of the beam are changed as the beam propagates at $z>z_0$ in such a way that the beam becomes a space soliton and, therefore, its J becomes equal to 2π. For example, let $2\pi<J<3\pi$. Show that I_0 increases and w decreases in this case. Since the power of the beam is constant and does not depend on z, then

$$I_0 w=Const \tag{15}$$

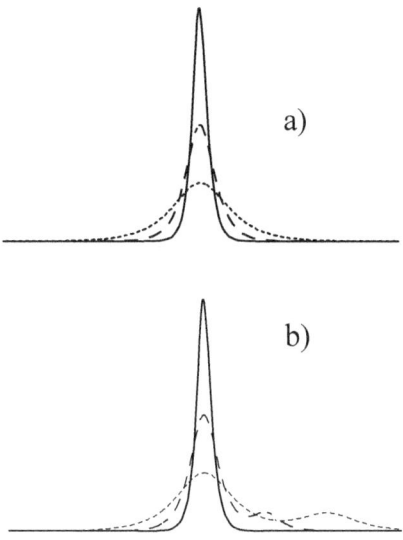

Fig.6. Transformation of profile of a light beam propagating in a nonlinear optical medium. (a) wide beam(dotted curve) approaches to the steady-state (solid curve); (b) two adjacent beams (dotted curve) approach to the steady state (solid curve).

Substituting w from (15) in (14), we have $J \sim I_0^{-1/2}$ and J decreases with an increase in I_0. If $J > 2\pi$, I_0 increases until J becomes equal to 2π. The same conclusion follows from (11)

If there is a background in a form of an additional light wave propagating along the z-axis in the region $-h < x < h$, $-\infty < y < \infty$, where $h \approx w$, then in accordance with the above consideration this background wave will be drawn in space soliton and width w of this result soliton decreases. If the energy of the background wave is restored repeatedly due to radiation of light by atoms excited at the gas discharge, we have a gradual increase in the light intensity within the space soliton.

Fig.6a shows how the structure of a light beam presented by line of points is transformed gradually to a steady state structure presented by solid line. The structure of a light beam is gradually narrowed, and its intensity increases. One of such structures is presented by dotted curve. As is seen in fig.6b two adjacent beams can be combined in one beam of total power.

Ought to underline that this situation differs radically from the situation at which the same background light wave propagates in a inhomogeneous linear optical medium where distribution of the refractive index is determined by the expression $n(x,y,z)=n_0+n_2I_0\mathrm{Ch}^{-2}(x/w)$, i.e. coincides completely with distribution $n(x,y,z)$ produced in an homogeneous nonlinear medium by the space soliton. By way of illustration, trajectories of light beams in the first case are shown in fig.7a. As is seen maximal deflections of light beams from the plane $x=0$ are constant as light beams propagate along the z-axis. The picture is invariant under a shift along the z-axis. Trajectories of light beams in the second case are shown in fig.7b. As is seen maximal deflections of light beams from the plane $x=0$ decreases as light beams propagate along the z-axis. But the deflections tend to a constant as $z\to\infty$. Unlike the first case where distribution of the refractive index $n(x,y,z)$ in space is constant and does not depend on the light waves, in the second case distribution of the refractive index $n(x,y,z)$ in space is variable and depends on the trajectories both the space soliton and background light beams. In other words, the background light beams increase the refractive index in regions where they propagates. As a result, they are concentrated around the region where n is maximal. If background waves are generated at all times along the space soliton, they are drawn in the soliton in the same manner and light intensity within the soliton increases progressively.

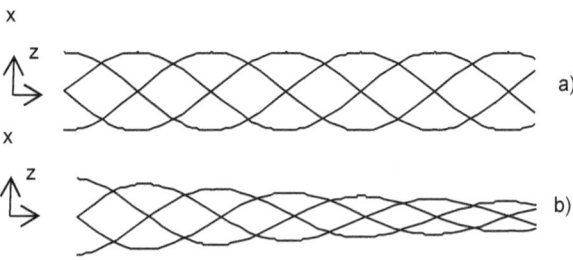

Fig.7. Trajectories of light beams in optical mediums. (a) inhomogeneous linear medium with the refractive index increased in the region where the beam propagates; (b) homogeneous nonlinear medium where an intense light beam propagates.

24

Now let show that the same effect is possible for a Ball Light that is a generalization of a plane optical incoherent space soliton. For this purpose we will try to convert the considered above situation with plane space soliton and plane background wave which both propagates along the z-axis into a situation where spherical space soliton is surrounded by plane background waves. Both light waves within the spherical soliton and background waves propagate in all possible directions. The conversion can be carried out in two steps. At the first step we convert the plane space soliton propagating along the z-axis into N plane waves of intensity $(I_0/N)\text{Ch}^{-2}(x/w)$ where i-th wave ($i=1, 2,..N$) propagates in parallel to the plane $x=0$ in the direction which makes an angle $\alpha_i=(2\pi/N)i$ with the z-axis. It is easy to check that the summary intensity of these waves in any point of the space is equal to the intensity produced by the initial plane space soliton. As a result, these N waves form the same distribution of n in the space and therefore the same plane lightguide where they propagate. Analogical conversion may be performed for the background wave. In a like manner, these N waves can draw in itself background light waves propagating in parallel to the plane $x=0$ in all possible directions.

At the second step let increase the curvature of the plane lightguide from zero to some definite value. In this case, the plane lightguide is transformed in Ball Light and the background waves are transformed in a majority of light waves propagating in parallel with planes tangent to the Ball Light surface. Such situation takes place at gas discharges where a majority of light waves produced by excited atoms are propagating in all possible directions and, in particular, in parallel with planes tangent to the Ball Light surface. In this case, the intense light circulating within Ball Light takes up a part of the energy from these waves.

Thus, appearing in gas discharge because of instability of an intense light propagating in a gas mixture, AO draws in its shell molecules of the gas component with maximal n. Besides, AO draws into its shell the light energy radiated by excited atoms within its shell and the light energy of background light waves propagating near its shell in parallel with the shell surface. Like an intense light provides an existence of Ball Light shell with increased n and Ball Light shell confines the intense light in the shell, the intense light circulating in the shell draws in the shell molecules of the gas mixture component with maximal n and the shell in turn draws in the background light surrounding the shell.

Let estimate the distance that is required to drawn in an adjacent beam within the plane space soliton. In accordance with the eikonal equation the additional shift of light beam in presence of a gradient of the refractive index is determined by the following expression

$$d^2\Delta x/dz^2 = g_n \tag{16}$$

where Δx is the deflection of the light beam from a rectilinear line, g_n is the gradient component perpendicular to the light beam. In our case $g_n \approx \Delta n/(w/2)$ where Δn is the maximal increase in the refractive index at $x=0$. Integrating (16) at initial conditions $\Delta x(0)=w/2$, $d\Delta x/dt=0$ at $t=0$, we have $\Delta x=w/2-g_n z^2/2$ and, therefore, $\Delta x=0$ at $z=(w/g_n)^{1/2}$. For example if $w=10$ μm, $\Delta n=10^{-3}$ then $z\approx450$ μm. This distance ought to be increased for Ball Light because (16) is transformed as follows

$$d^2\Delta x/dz^2 = g_n - R^{-1} \tag{17}$$

where R is Ball Light radius. In the above example $g_n=2000$ m^{-1}. If $R=0.5$ cm then $R^{-1}=200$ m^{-1} and influence of the curvature is insignificant.

One can see that a Ball Light is a very good accumulator of light energy. Process of accumulation of the light energy terminates if the energy sucked into Ball Light per one period of light rotation is equal to the losses in the same time. If conditions for suction are preserving for a long time, the density of the light energy within Ball Light may be significantly greater than that in the space surrounding Ball Light. Ought to note that the losses within Ball Light (both irradiative and because of molecular light scattering) decrease with an increase in the light intensity within Ball Light. In this case, a situation is possible where the light energy introduced into Ball Light per one period of circulation is greater than the total energy dissipated at the same time and a gradual increase in the light energy can take place. Possibly, this effect takes place for natural BLs too. In this case, a usual sunlight is used as background light waves that are drawn in the BL. There are many evidences that BLs are observed in a sun day when no thunderstorms are observed. A storage of the light energy is an extremely difficult and in the same time extremely important problem. The light energy can be stored in optical resonators. But time of the storage is about several microseconds in the best case. As follows from observations of BLs, the time of the storage within BLs can be several minutes at least.

Production of an intensive circulating light by exothermal chemical reactions

There were reports of new types of autonomous objects on the border 20 and 21st centuries. Unlike previous luminous autonomous objects (AO) whose lifetime was a fraction of a second, the lifetime of the new types reached several seconds. The first report of such objects appeared in 1997 [Emelin 1997]. A photo of bouncing glowing balls (GBs) in a form of their continuous trajectory obtained in a dark with an open shutter of photo camera is shown in Fig. 9. Bouncing glowing balls of a few millimeters size are observed, which had abnormally high jumping ability. Balls disappeared without a trace at any point of its trajectory. GBs can make on the surface of an ordinary table about 50 jumps. Since there is a horizontal component of GB velocity in the photo, the GB trajectory is a set of clearly defined parabolas. Each parabola corresponds to one jump. There are 20 parabolas in the photo. Lifting and dropping branches of parabolas are identical in their durations and therefore the time of lifting t_{LIFT} and the time of dropping t_{DROP} are identical

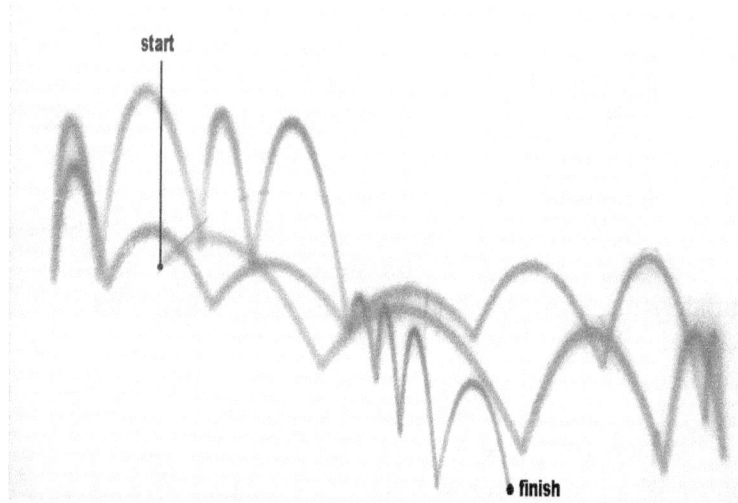

Fig.9 Trajectory of bouncing ball

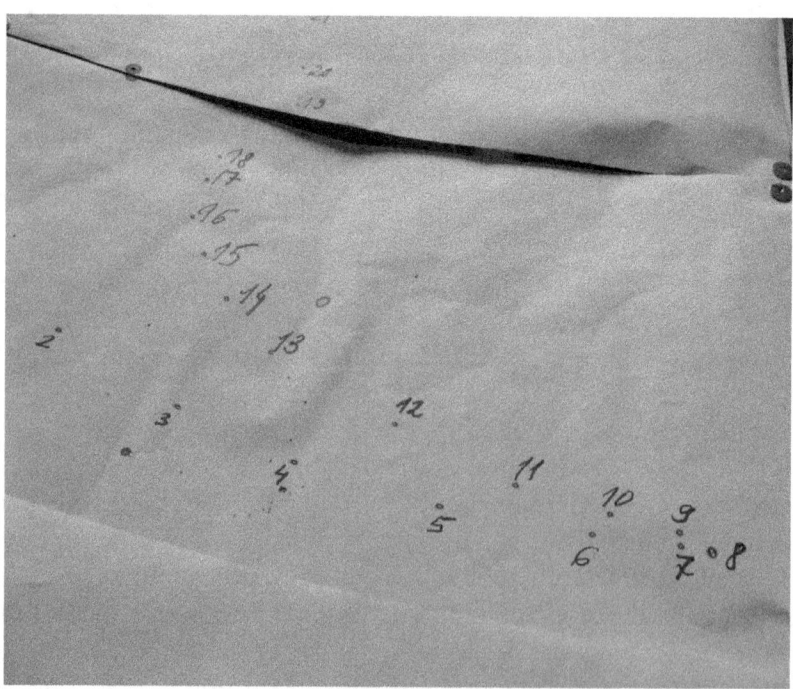

Fig. 10 Trace of the bouncing ball sheets of fax paper. Number near the spot shows a number of the jump.

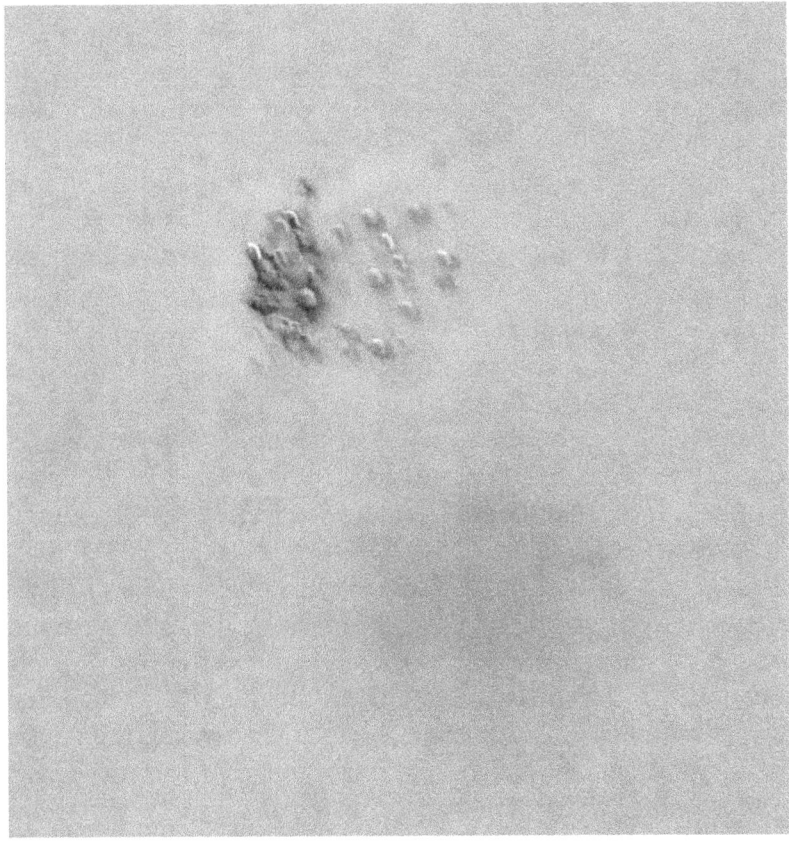

Fig. 11. Bouncing Ball that is separated at the majority of small balls at a strike against the horizontal windowpane placed on the table. The dark spot below the ball is the trace of initial heating of the surface of the table through the glass.

For the sake of justice ought to note that an analysis of the bouncing ball in detail was presented by Stakhanov in case number120. A glowing ball about 4 mm diameter raised at short circuits. The ball made 12 jumps for 4 second on the rubberized tape of the conveyor. The height of jumps was about 1.5 cm. The ball diameter was decreasing gradually till 1 mm and then the ball disappeared without leaving trace. Stakhanov calculated the time interval between adjacent jumps. It turned out that the time was greater by 3 times than the time of a solid ball in the field of gravity. Stakhanov concluded that the density of the ball is close to the density of the air.

Another group of experiences is connected with attempts to produce AOs at the powerful gas discharges. Notions about a ball lightning presented by Abrahamson and Dennis [Abrahamson 2000] are used. According to these representations, ball lightning exists owing to oxidation of silicon nano particles in an atmosphere. Such particles are formed at the stroke of a usual linear lightning in the ground as a result of reaction in thickness of the ground of oxides of silicon and carbon. Though these representatives cannot explain the facts of BL movement against a wind, penetrations through window panes, catching up airplanes, however, they promote some successes at attempts to product artificial BL. Really, as follows from these representations, an erosive gas discharge is required at which additional products evaporated from electrodes should be delivered in the BL region.

Brazil scientists reported in 2007 that glowing balls (GB) can appear at conventional arc discharge with small pieces of Si wafer introduced in the discharge gap [Paiva 2007]. Unlike conventional sparks that accompany an arc discharge, GBs have unusual properties. In first, their lifetime is essentially greater than that of the sparks and can achieve eight seconds. In second, they jump in process of their moving, can bypass obstacles, and penetrate through splits which width is smaller than their diameter. In opinion of the scientists, these GB properties remind behavior of Ball Lightning. Video film about the behavior of these objects can be seen on a site http://www.youtube.com/watch?v=fsu8IaaVsvk

These experiences have been repeated by the Holland researchers. Their video film is presented on a site http://www.youtube.com/watch?v=QLTPELhKAYM andhttp://www.youtube.com/watch?v=QLTPELhKAYM&mode=related&search .

An electronic paper about the results of their experiments entitled "Artificial Ball Lightning produced in the laboratory!?" is presented on the site

http://members.chello.nl/r.dekker49/boBall Lightliksem/boBall Lightliksem.htm

There are 2 video films from viewing of which it is possible to be convinced that lifetime of some luminous balls is about 6-8 seconds and that they jump aside from a table more than 10 times.

The height of jumps varies, and the height of the subsequent jump can be greater than that of previous one. Similar film is presented by the American researchers on a site
http://www.youtube.com/watch?v=T7fUKEGyxS8&NR=1and
Spanish investigators on site
http://www.youtube.com/watch?v=IYSie5YsaOI .

Thus, now foreign researchers have reached that level which was shown by the Russian researchers 5-10 years ago. As can be seen, foreign researchers should investigate more many still that have been investigated in Russia already.

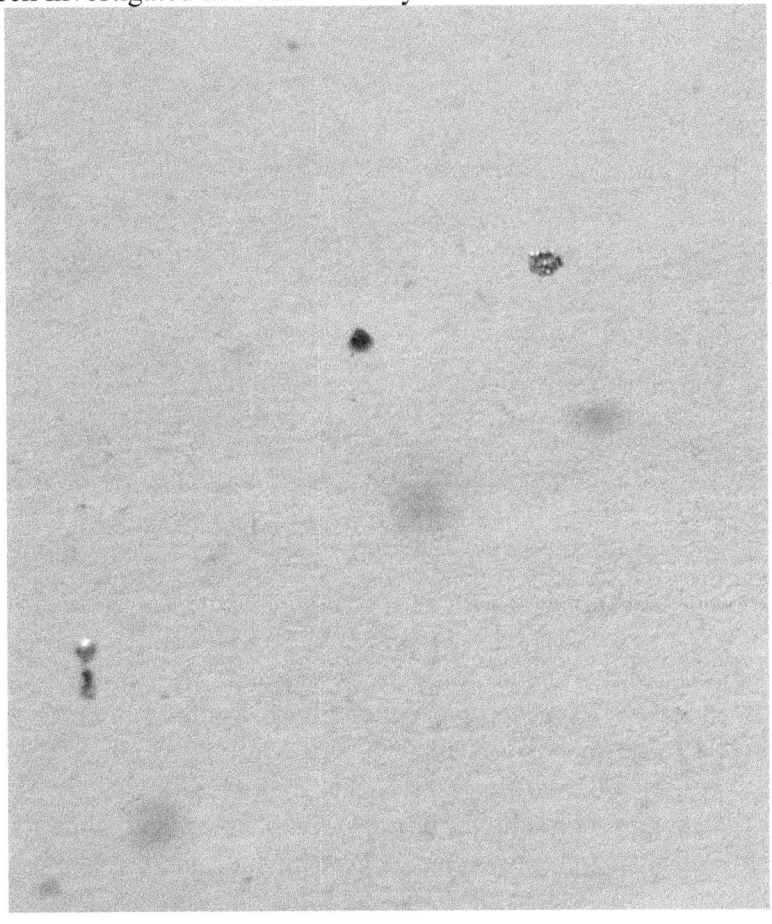

Fig. 12 Traces of balls on the place surface of glass of 4 mm thickness. As is seen, there is no bouncing.

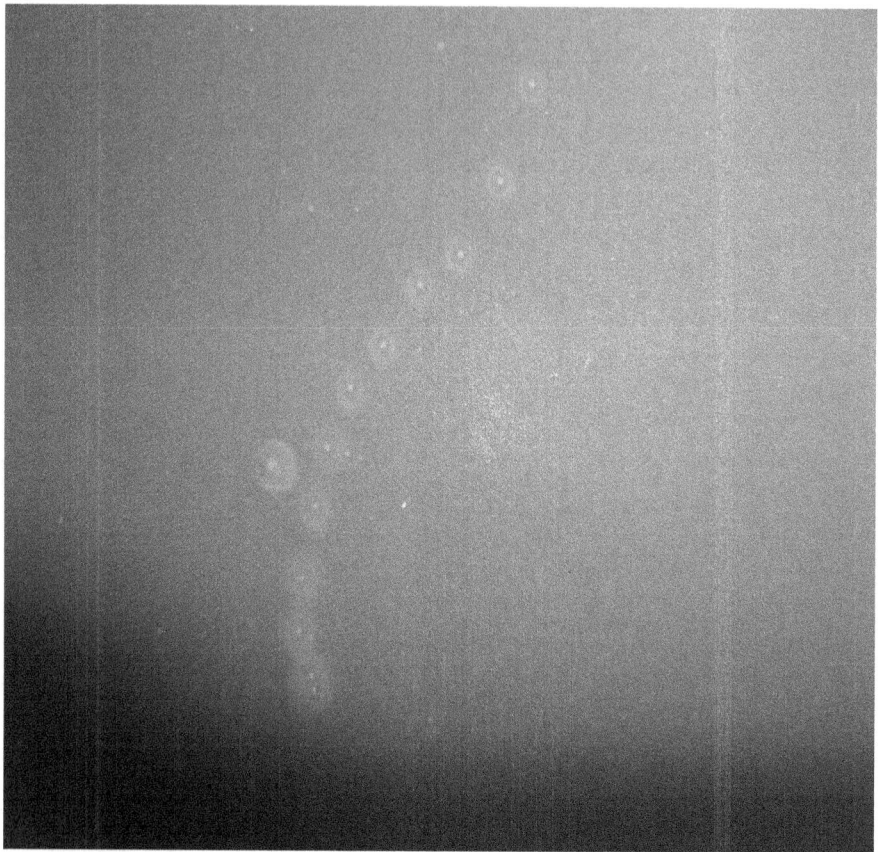

Fig. 13 Trace of the bouncing ball on the black paperboard

The most informative is video by V. Gooses, P. de Graaf and R. Dekker from Holland, Eidhoven **Ошибка! Источник ссылки не найден..** They accompany their video by description in detail of their experimental installation and features of experiment. They also think that their GB is a certain form of Ball Lightning and present the most known at present theory of Ball Lightning of A. Abrahamson and J. Dinniss 0. But the question whether similar GB is a certain version of Ball Lightning remains open. It is a good chance to apply our theory to resoles this problem.

One can see in the video presented in **Ошибка! Источник ссылки не найден.** that the GB can jump on the surface of the white table during 8 s. The maximal height of the GB is about 10-

20 cm and the maximal height decreases gradually with time. Nevertheless there is the jump maximal height of which is greater than that of the previous jump (second attempts, third jumps in sixth second in [Gooses 2008]).

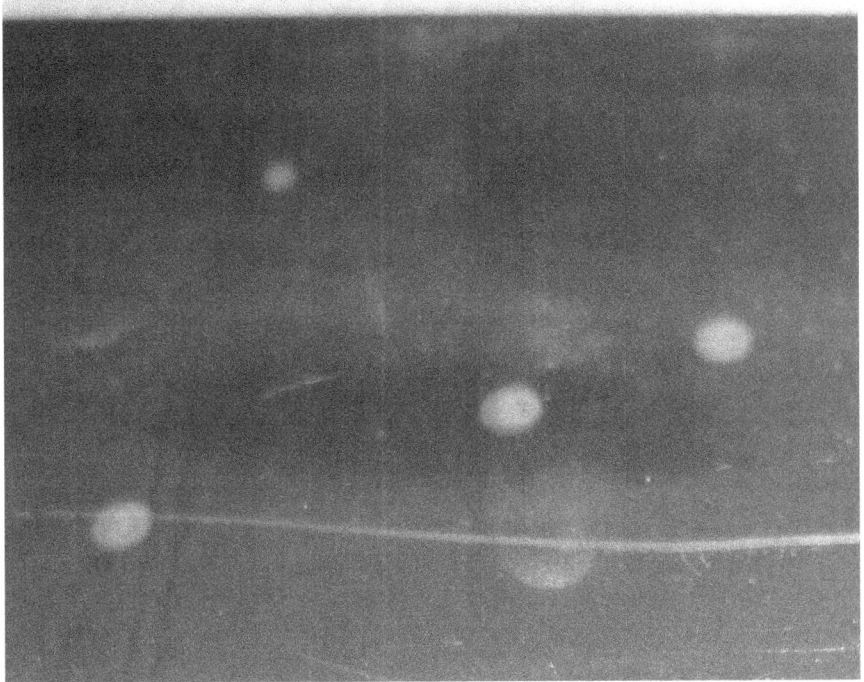

Fig. 14. Trace that the bouncing ball leaved at tantalite at bouncing. The small ball is seen in the top right-hand corner. Possibly, the light energy within the ball becomes insufficient for next jump

Besides, we carried out some additional experimental investigations to clarify a physical nature of GB. Our installation is described in detail in [Torchigin 2010]. The piece of the silicon wafer of 0.4 mm thickness and 2*2 mm2 area is placed between two wolfram electrodes that are connected with the battery of automobile accumulators of 36-48-60 through switcher that capable to switch up to 300 A current. Initially the current through the silicon wafer is absent because the wafer is dielectric. The initial voltage pulse applied to the electrodes breaks down the dielectric. Then the switcher is turned on. As a result, the silicon wafer heats up and melted.

Drops of the liquid silicon in a form of GB fall on the horizontal surface of the table. The further behavior of the drops is identical to that in the mentioned above experiments.

Fig. 15. Photo of the ball as compared with the ball in the ball-point pen. The surface of the ball is brilliant. It is supposed that the ball consists of the silicon and its surface is covered by silica.

It turns out that the study of the character of the behavior of the drops on various surfaces can give much additional information about physical processes that takes place at the interaction between the GB and the surface. Fig. 10 shows traces of GBs at impact with the surface in a form of the thermal paper used in faxes. The fax paper has the following properties. The paper becomes dark at heating above 80^0 C. If the face of a hot circle rod heated at the temperature smaller than 250^0 C contacts with the paper surface, a trace in a form a dark circle remains on the paper. If the rod temperature is greater than 250^0 C, the trace in a form of a white circle surrounded by the dark ring remains on the paper. An

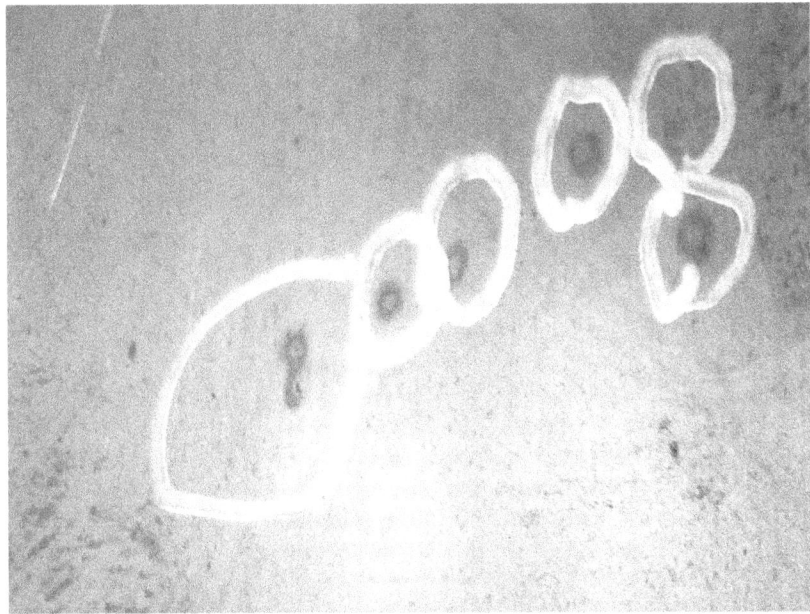

Fig. 16. Trace of the bouncing ball on the foil getinaks. Each trace is rounded by the circle made with felt-tip pen.

appearance of the dark ring is explained by propagation of heat along radial directions with the gradual decrease of the temperature. The darkness of the ring decreases with an increase of the distance to the center because the temperature of the paper decreases also. The area where temperature is in the range 80^0 C<T<250^0 C occurs dark.

Sequential collisions of GB with fax paper are numbered by hand in Fig.10. As is seen, there is about 20 GB jumps on the fax paper. Thus, there are no doubts that the GB heats the obstacle at collision. It is not clear in advance either a direct contact with obstacle takes place or the paper heating takes place due to light radiation at GB approaching to the obstacle.

To clear up the process of GB colliding with the surface of the obstacle the following experiment was carried out. A glass plate from windowpane of 4 mm thickness was used as the obstacle for GB. The glass plate has property to pass the radiation through itself. In other words, the windowpane is not heated by radiation

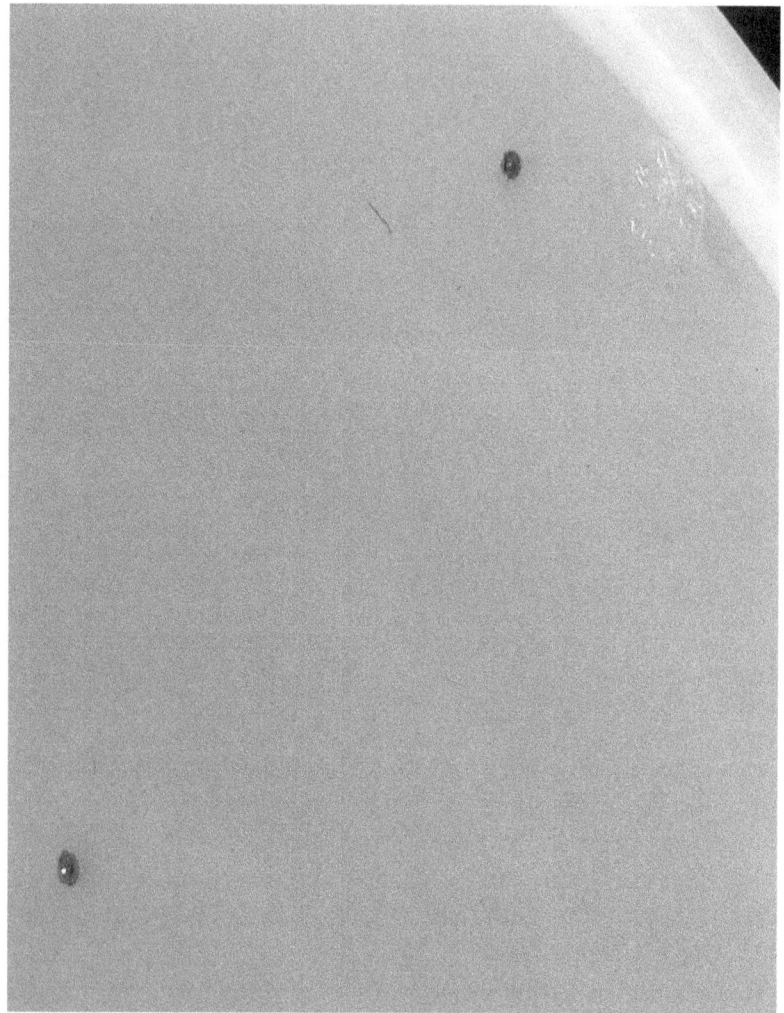

Fig. 17 Photo of the bouncing balls on the plastic vessel filled by water. The traces of burning of the bottom are seen near the balls.

noticeably because the radiation penetrates through the glass. In this case, the GB hits the glass surface and is separated into a set of small balls as is shown in Fig.11. As is seen, GB damages a little a fragment of the smooth top surface of the glass plate. The dark spot is seen on the white paper under the glass plate. Bouncing GBs on the glass surface were never observed. As is seen in Fig. 12, ball are located near the traces where they collide with the glass plate.

Traces of bouncing GB on the black sheet of paper are shown in Fig. 13. One can see that there is the small white spot is surrounded by the circle white spot of greater diameter. In is explained by the fact that the GB bouncing in all experiments is accompanied by the white smoke that is produced by the GB in all time of its existence. This is valid for the Emelin *et al* experiments, [Emelin 1997]for the Paiva *et al* experiments [Paiva 2007] for the V. Gooses, P. de Graaf and R. Dekker experiments [Gooses 2008] as well as for our experiments [Torchigin 2010]. The same white traces that the bouncing GB leaved on the tantalite plate are seen in Fig. 14. The small ball is seen in the top right-hand of the figure. This is the last contact of GB with the tantalite. As is seen in Fig.13 and Fig. 14, the distances between adjacent spots are not identical.

The ball that is seen in the top right-hand of Fig. 14 is shown in Fig. 15. One can see that the surface of the ball is brilliant.

A behavior of GB on the surface of thin foil bounded to getinaks is shown in Fig. 16. The traces of the GB are faintly visible. The traces are rounded by circles made with the white felt-tip pen. Unlike the glass plate, several jumps are seen.

A behavior of GB on the surface of water has been studied also. It turns out that GB slides on the surface without decreasing its brightness in several second. As a rule, the glowing ceases with GB disappearance. The events are observed when the sliding GB jumps on the surface once. Sometimes the glowing GB penetrates the water. Two GBs on the bottom of the plastic vessel is shown in Fig. 17. As is seen, there is dark spot under the ball. The spots arise due to heating the plastic. Thus, the glowing penetrated through the water. The thickness of the water layer is about 1 cm. The video is presented in [Torchigin 2010] where GB is glowing on the bottom of the vessel filled by the water.

For the sake of justice, we need to note that we have observed many glowing hot balls properties of which coincide with that of the liquid drop heated at high temperature. These balls are crumble into small fragments in a collision with any obstacle. The color of these ball changes gradually from white to yellow, orange, red, dark red. The same takes place with any conventional body.

We can mark the following obvious anomalies

1. The brightness of the GB is not changed in time unlike the brightness of an incandescent body heated at very high temperature at which the body is cooling down gradually. In this case, the white color of the body becomes gradually yellow, orange, red, dark red and so on.

2. There is an abnormal great number of jumping from the surface of the fax paper. A conventional small ball made off any known material cannot jump on the paper.

3. There is smoky-colored trace after bouncing GB.

4. GB heats up the surface at a collision with it.

5. A number of jumps on the surface depends on properties of the surface. The maximal number corresponds to the surface that is heated maximally by the GB. This is a thin black leaf of paper. There are no jumps on the solid metal surface or on the transparent glass surface. A great GB that collides with the horizontal transparent windowpane of 4 mm thickness is crumpled into small fragments. A small GB stops near the region of colliding.

Simple estimates and experiments show that the solid ball of a few millimeters heated to white-hot temperature of about 2000C, cools quite rapidly within a few seconds. Furthermore, the body varies its color during cooling from white to yellow, orange, red, etc. We have the video of such balls. These balls made no jumps at colliding with the paper. However, as can be seen from the videos in Youtube, glowing balls can shine without any significant decrease in brightness and color for at least six seconds.

Let us assume that silicon within the GB is in a liquid state. In this case, the total heat stored in the GB increases significantly due to great specific melting heat of silicon. Let initial GB temperature is $T_0=2000^0C$. At GB cooling the following energy is released

$$Q=m\,(Q_1+Q_2+Q_3)$$
where $Q_1=(T_0-T_{melt})c_0,$

$Q_2=c_1,$

$Q_3=(T_{melt}-T_1)c_2,$

$c_0 = 20.1$ Jmol^{-1}K^{-1} =718 Jkg^{-1}K^{-1}– is the specific heat of liquid silicon,

c_1 =49.8 kJmol^{-1}=1780 kJ kg^{-1}is the specific melting heat of silicon,

c_2 =20.1 Jmol^{-1}K^{-1} =718 Jkg^{-1}K·isthe specific heat of solid silicon,

T_{melt} =1415^{0}Cis the melting temperature of silicon,

T_1=500^{0}Cisthetemperatureatwhichradiationfrom GB is very small as compared with initial one at the temperature T_0=2000^{0}C.

For GB parameters presented in the above example the GB energy decreases at decreasing its temperature from T_0=2000^0 C to T_1=500^0 C at Q=1.019 10^{-6} J.

On the other hand, the power radiated by the GB is determined by the following Stefan-Boltzmann relation

$$W_r = \sigma T^4 \tag{18}$$

For absolute black body σ=5.67 10^{-8}W m^{-2}K^{-4}. Even at the lowest temperature T_1=5000C the GB radiates power W_r=126 10^{-6} W. A time interval in which the GB losses all its stored energy at heating from 500 to 20000C is smaller than the limiting time T_{lim}=Q/W_r=8ms.

Thus, even assuming that all time the GB radiates the minimal power W_r which is smaller by 65000 times than maximal one the energy stored in the GB vanishes faster in the time interval that is smaller than 8 ms. A hollow GB of greater diameter losses its energy even in smaller time. Some impression about time of a glowing of similar objects can be obtained from sparks that are taking off from a rotating grinding stone. These sparks are the hot grains that have separated from the grinding stone. Their lifetime is a fraction of a second. Such grains cool down gradually, therefore their brightness weakens gradually. Change of luminosity of jumping GBs in the video film during their lifetime (more than 5 s) is imperceptible.

GBs in the video file are shone without appreciable change of light intensity within several seconds that is longer by thousand times. An explanation of this phenomenon can be the following. A careful viewing of the video we can see that the GB that are rebounded from the floor or table surface are accompanied by smoky trail. In addition, all GB at the end of his life disappear without a trace. We can therefore assume that some exothermal reaction is responsible for the prolonged SS glow. This reaction maintains a high temperature at which a sufficiently intense light is emitted, a portion of which remains in the GB shell. The same is valid for the sparks from the fire that radiate the light which brightness is constant. These sparks are also extinguished without leaving a trace, after completely burn (oxidize as a result of a chemical reaction). Since the GBs are the result of an electrical discharge in a silicon wafer, it can be assumed that the smoke trail left by the GB is a product of the reaction of silicon with atmospheric nitrogen. It is known from chemistry that the product of the reaction between silicon and nitrogen is a gray powder. The gray powder is clearly seen in all video with bouncing balls. Moreover, as is marked in [Emelin 1997], when GB is directed in forevacuum, no glowing ball observed. Indeed, there is no oxygen and nitrogen and, therefore, no exothermal reaction is possible.

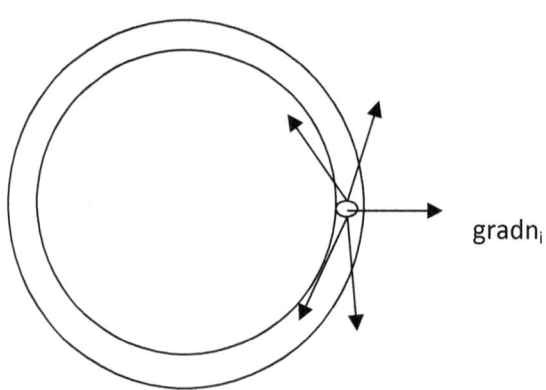

gradn$_i$

Fig. 18. Capture of the light radiation radiated by a hot atom located in the transparent liquid shell of bouncing ball

Thus, the nature prompts us one more way for the introduction of the light radiation in the BL transparent shell. This is a direct radiation of hot atoms that are capable to radiate photons. Let us determine a part of the radiation that is confined by the shell. The refractive index of the shell is about $n=1.45$ because the shell is the liquid glass produced by the oxidation of the silicon at heating (burning the silicon). All beams radiated by the atom that are subjected by the total inner reflection cannot penetrate in free space and rotates inside the shell.

Let us determine the space angle where the beams radiated by the atom are circulated in the shell. The angle of the total inner reflection is given by $\alpha=\mathrm{ArcSin}(1/n)$. At $n=1.45$ we have that $\alpha \approx \pi/4$. All beams that make with the perpendicular to the side surface of the shell the angle that is smaller than α are radiated in free space (Fig.19). These beams are located in the space angle equal to $2\int_0^{\pi/4} 2\pi R Sin\alpha(Rd\alpha) = 4\pi R^2(1-1/\sqrt{2})$. Therefore, only about 30% of light beams are radiated in free space and 70% is confined by the shell.

Seemingly, 70% of all radiation radiated by the atoms should be accumulated in the shell. However, there are inevitable dissipative losses in the shell that can be estimated as follows. As is known, the quality factor Q of the micro-resonator of the whispering gallery light wave in a form of a glass ball is about $Q=10^{10}$. This means that the time constant of decreasing of the intensity in the resonator is relatively great and is given by $T=Q\tau$ where τ is the period of oscillation of the light wave. On assumption that $\tau=\lambda_0/c$, where $\lambda_0=0.5$ mµ we have $T=1.6\ 10^{-5}$ s.

The same order of the magnitude can be determined from the following consideration. The losses of the conventional fiber used in the telecommunications are about 0.2 Db/km. The radiation decreases by two times when the losses are equal to 3 Db or at the distance $L=15$ km. The light propagates 15 km at the time $T=L/(c/n)= 0.75\ 10^{-5}$ s. As is seen the estimations differ by two times only.

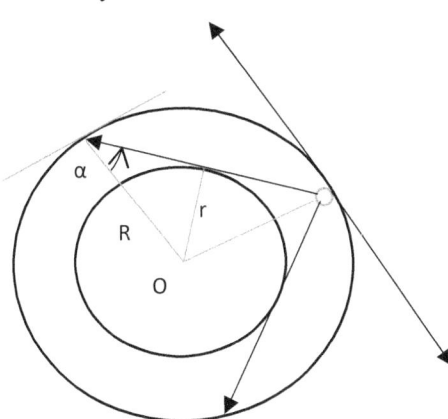

Fig.19. All beams radiated by a hot atom located on the surface of a transparent ball in the ring between the circumferences of radius r and R are circulated within the ball

Let us now determine the power that is produced at burning the bouncing ball. Let the radius of the ball be $r=1$ mm and, therefore, its volume $V=4$ 10^{-9} m^3. Since the density of the silicon is equal to $\rho=4.6\ 10^3$ kg m^{-3}, the mass of the ball is equal to $m=19.2\ 10^{-6}$ kg. The total energy produced by the exothermal reaction is given by $E=Am$ where A is the heat of the combustion (about 50 MJ kg^{-1}). The duration T_B of the time when the ball is burning is 8s. Then the power $P=Am/T_B=120$

Watt. The power in the shell is smaller by 70% and is equal to 84 Watt. However, this power can be accumulated at $T=0.16 \cdot 10^{-4}$ sonly. Then the energy stored in the ball is equal $E=PT=1.3 \cdot 10^{-3}$ J. The density of the energy is given by $W=3E/(4\pi r^3 h)$ where h is the width of the shell. On assumption that $h=10$ μm, we have $W=0.3 \cdot 10^{11}$ J/m³.

Thus, the great time of GB luminescence could be explained on assumption that there is some chemical exothermal reaction that supplies heat energy to the GB. Similar situation takes place in small sparks lifting from open firewood. The sparks radiate red light in several seconds due to oxidation of small wood particles. The same thought is expressed in the following commentary to the video file [Gooses 2008]. "I don't think this is ball lightning. It is tiny balls of burning silicon. Silicon burns very hot (like magnesium) and is light weight. They get lighter and lighter as they burn up, causing them to bounce higher. They also eject material in all directions as they burn. Material ejecting downward pushes them off the counter top".

Possibly, everything is correct in this commentary besides a single exception. The material ejecting downward is not able to push balls off the obstacle. There are no known physical phenomena that provide appearance forces applied to the ball and directed upwards because the ball ejects material in all directions. Moreover, as is seen from GB behavior at colliding with the windowpane the ejection downward of silicon particles does not "push them off the counter top".

Thus, the class of ball lights is much broader than it might have been expected initially. Ball light shell may be formed not only by compressed air or by the components of the gas mixture the reflection index of which is greater than that of the air. The Ball Light shell may be formed with a transparent liquid, such as silicon dioxide. As is known, the silicon dioxide is an excellent optical medium used for the production of fiber optics. Thus, we have derived one more method of production of the intense light that is circulating within the Ball Light shell. The light in the shell can be generated not only at the time of the electric discharge when Ball Light rises. The light in the shell can be introduced into the shell gradually when Ball Light shines due to the exothermic chemical reaction. In this case, the lifetime of Ball Light is determined by a duration of the chemical reaction. The lifetime of Ball Light is equal to the duration of the chemical reaction. The reaction is terminated when the Ball Light size becomes equal to zero. The reaction is terminated with the Ball Light disappearance. The Ball Light disappears completely. We can see that no electrical discharge is required for production of Ball Light of this case. It is sufficient to heat up a piece of silicon at the high temperature at which the piece is melted and a ball is formed. The further increase in the temperature exerts the chemical reaction that entails a complete oxidation and nitrogenization of the silicon.

It is important that the light radiation from the GB does not heat the surrounding air directly. A nontransparent object is required that absorbs the light radiation. The light energy is transformed into the heat energy in such object. This entails an increase of the temperature on the object surface. In turn, the temperature of the air layer contacted with the object surface increases too due to phenomena of heat conductivity. Thus, a nontransparent object is required to provide a gradient of the air refractive index. This is the reason why a GB tends to bypass an obstacle. Ball lightning bypasses obstacles owning the same reason [Torchigin 2003].

Consider qualitatively processes in the gap between the GB and a plane obstacle at GB approaching at the distance when the gap width is smaller than the GB radius (Fig.20). In any case, the surface of the obstacle is heated up. The heating depends on the material of the obstacle. Heating the obstacle surface depends on its specific heat capacity and specific heat conductivity. The greater these parameters the smaller the heating is. At last, the heating depends on thickness of the obstacle. Thin sheets in the form of a foil heat up faster by

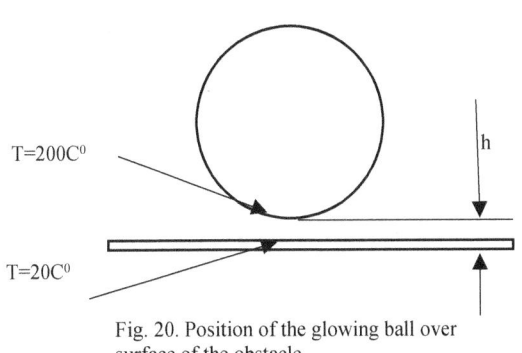

Fig. 20. Position of the glowing ball over surface of the obstacle

several orders of magnitude than a flat surface of a continuous material. In the latter case, the heat spreads in depth of the material and the thermal resistance of a continuous material much smaller than that of a thin sheet. The thermal resistance of a thin sheet is in inverse proportion to its thickness. In this respect a paper sheet is a rather suitable material for fast heating (heat capacity of cupper is 380 J kg^{-1}K^{-1}, heat capacity of paper is 1.51 J kg^{-1}K^{-1} that is smaller by 250 times, specific heat conduction of cupper is 401 W m^{-1}K^{-1}, specific heat conduction of paper is 96 10^{-3}W m^{-1}K^{-1}that is smaller by 4171 times).

The hot surface of an obstacle heats layers of the air located near the surface owning the heat conduction of the air contacted with the surface. It would seem, at heating air its pressure increases and there should be the forces directed upwards. However, these forces are very small and can be neglected. Indeed, the GB speed is about 0.2 ms^{-1}. In the same time, the sound speed in air is 340 m s^{-1} that is greater by 3 orders of magnitude. Because of this, transient processes connected with a change of air pressure proceeds faster by 3 orders of magnitude than GB position changes. Any insignificant increase in the air pressure quickly enough resolves.

This mechanism could work if GB was disposed inwardly cylinder, which diameter is equal to the BL a diameter. In this case, the air heating would lead to an increase of the air pressure that would lead to an occurrence of additional force applied to the BL from the side of the hot compressed air. In this case, the lateral wall of the cylinder prevents the expansion of the hot gas in all sides. This method of transformation of the thermal energy in the mechanical one is used in the internal combustion engine where the air is heated up not from a hot cylinder, but at the expense of fuel combustion. However, in the case under consideration, the lateral walls of the cylinder are absent and the appreciable excess of pressure upon the GB bottom surface is not created.

Thus, we can conclude that the heating the surface of the obstacle under the GB promotes to a great number of GB jumps. However, as is known from mechanics, the temperature cannot provide the necessary mechanical momentum to GB. The force is required that is directed upwards and that acts a certain time. Usually when a conventional steel ball is jumping on a marble horizontal surface there is the vertical force applied to the ball. The force acts from the side of the marble surface. The force provides a change of the vertical velocity of the ball. Two necessary conditions are required to satisfy that the ball jumped up to the same height from which it fell. First, the ball should be solid and elastic. For example, the ball made of clay, wood or chalk cannot jump. Second, the surface should be also solid and elastic. In this case, the elastic deformations of the ball and surface take place. As a result, there is the force that is applied to the ball from side of the surface.

At least, the second condition is not fulfilled for the fax paper. No jumps is possible on the thermal fax paper for any ball. However, a number of GB jumps is great. Moreover, GB is not jumping on a solid and elastic transparent windowpane. The question arises. What does reason cause the ball to change the direction of its velocity? There is the hot air only. In is impossible to imagine that the hot air can change GB velocity.

It is necessary to find other physical phenomena leading to occurrence of forces that change a direction of the GB speed after the collision with the fax paper. We will consider one of such phenomena. Usually attention is not paid to the phenomenon though it is based on the generally accepted physical laws. For this reason, the consideration will be presented in detail. As was shown, the density of the optically induced force applied to the inhomogeneous optical medium where the light is propagating is given by

$$f = -grad(\varepsilon)W_L \tag{19}$$

The gradient of the permittivity is not constant on the GB shell. The gradient is maximal in the region where the distance between the surface of the obstacle and the GB surface is minimal. Since the fax paper becomes black at the temperature about $T_{PAPER}=200$ C^0, let us assume that a difference of temperatures within the region is equal to $\Delta T = T_{PAPER}-T_0$ where $T_0 =20C^0$ is the temperature at normal condition. Since the permittivity of the air is in the inverse proportion with the temperature, we have that the permittivity of the air that touches the obstacle is smaller than the permittivity of the air that touches the bottom surface of the GB by a factor $(273+T_0)/(273+T_{PAPER})=0.62$. In this case a difference between the permittivities is equal to

$$\Delta\varepsilon = \Delta\varepsilon_0 (273+T_0)/(273+T_{PAPER}) \tag{20}$$

where $\Delta\varepsilon_0=\varepsilon-1=5.4\ 10^{-4}$, ε is the permittivity of the air at normal conditions. Thus, $\Delta\varepsilon=5.4\ 10^{-4}\ 0.62=0.33\ 10^{-4}$. On assumption that the gradient of the air permittivity in the gap between the surface of the obstacle and the bottom surface of the GL is constant, we have

$$grad\ (\varepsilon) = (\Delta\varepsilon_0/h)(273+T_0)/(273+T_{PAPER}) \tag{21}$$

where h is the width of the gap.

Let us next consider the process of the interaction between GB and the obstacle. The left GB in Fig. 14 is subject to gravity and drops on the obstacle. The core of GB is the liquid silicon that is surrounded by the liquid silica SiO_2 that forms the shell where the intensive white light is circulating. The evanescent field of the white light penetrates through the boundary between the silica and surrounding air and is decreased exponentially with the distance from the boundary d as $Exp(-d/L)$. Length L depends on the waveguide mode in the shell and tends to infinity when the angle of incidence of light beam within the shell tends to the angle of the total inner reflection. Length L is comparable with the wavelength in free space for light beams that propagate in the shell along its surface.

The evanescent field of the white light in accordance with Eq. (3) compresses the air surrounding the shell. The layer of the compressed air is shown in Fig. 21 in a form of the external ring. At approaching the surface of the obstacle where the gradient of the optical medium (in a form

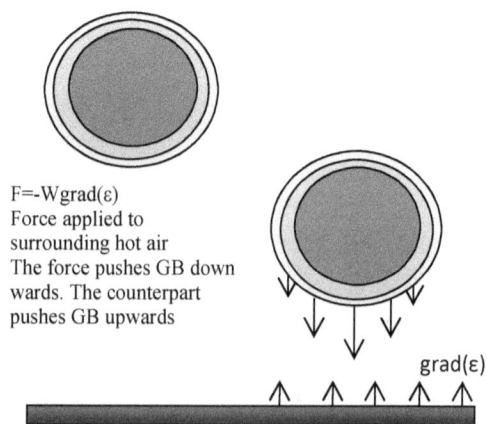

F=-Wgrad(ε)
Force applied to surrounding hot air
The force pushes GB down wards. The counterpart pushes GB upwards

grad(ε)

Fig. 21. Process of bouncing.

of hot air) is different from zero and is directed upwards, the Maxwell-like OIF arises in accordance with Eq. (19). The force F is applied to the optical medium (hot air) and is directed opposite to the gradient. The force is produced by the electrical field in the external layer of the compressed air where the density of the light energy is different from zero. In accordance with the third Newton law, there is the counterpart of force F. The counterpart is opposite directed and is applied to the object that is responsible for a rise of force F. This object is the external layer of compressed air. Since the layer is connected directly with the whole GB, the force $-F$ is applied to GB. If force $-F$ is insufficient to overcome the weight of the ball for given width d=0.1 mm, the ball approaches the surface and d decreases. In this case grad(ε) increases, OIF increases and it overcomes the gravity.

The question arises. How can GB repel from the conventional air? The answer is the following. There is the same phenomenon that is used in turbo reactive aircraft. The momentum of the air that enters the engine is smaller than the momentum of the air that leaves the engine because the velocity of the output air is greater than that of input one. As a result, the engine provides the momentum to the air directed backwards. In accordance with the third Newton law, the same engine provides the momentum to the engine directed forward. An increase of the velocity of the output air in the engine is provided by means of the air heating. The heating is provided by the incinerations fuel.

The increase of the velocity of the air in the gap between the surface and the bottom side surface of the GB is provided by the Maxwell-like force F=$-W$grad(ε) applied to surrounding hot air in the gap and shown in Fig. 21 with arrows directed downward. As a result, the air moves downwards. In accordance with the third Newton law, the counterpart of the Maxwell-like force provides the GB motion upwards.

Then the gradient of the permittivity in the gap is given by grad(ε)=$\Delta\varepsilon/d$=2 m^{-1}. In accordance with Eq. (19) the density force is given by

$$f=\text{grad}(\varepsilon)\, W = 6\ 10^{10}\ \text{N/m}^3. \qquad (22)$$

Thus, the density of the optically induced force can be greater by six orders of magnitude than the density of the gravity force.

This gives a hope that OIF can overcome the GB weight. However, the OIF density forces are applied to a small part of the ball volume that is located near the surface of the ball where the gradient is great, whereas the gravity is applied to whole volume of the GB. We deliberately presented rough calculations because an error of several orders does not matter on the final result that OIF is sufficiently great to change a direction of the GB velocity when GB drops on the obstacle. Solution that is more exact can be obtained on the base of investigation mathematical model of this phenomenon. We would like to remind only that, as was shown, values of physical parameters of the obstacle surface are varied by the several orders of magnitude. The same is valid for parameters of GB.

Besides, we did not take into account delays of the inertial processes connected with heating the obstacle by GB due to radiation and heating the air within the gap due to the heat conductivity. When GB begins to near an obstacle, the obstacle surface has no time to increase its temperature up to maximal one. The same is valid for the air layer contacted with obstacle surface. Even after the GB has stopped and has started to move in the opposite direction an obstacle continue to heat up the air layer. At the moment when GB stops, the obstacle heats up the air layer at the maximal speed. Thus, when GB is moving in the opposite direction, it is moving in the same space but gradient of the air refractive index in this time is greater than that when the GB moves towards to obstacle. In accordance with (19), the force that accelerates GB is greater than that that breaks it. Besides, the thickness of the air layer with noticeable gradient of the refractive index is increasing in time. The GB is accelerated in opposite direction not only by greater force but also at the greater distance. Because of this, a positive work that is made over the GB in accelerating process can surpass considerably the negative work spent for GB braking. In other words, the kinetic energy acquired by the GB in process of acceleration can surpass significantly the kinetic energy loosed by the GB in breaking process.

Taking into account the consideration, it is not so hard to find out the reason why the next jump can be higher as compared with previous one. If GB contacts occasionally with the place of the obstacle where a small deepening takes place, conditions for air heating in the gap between the bottom GB surface and the deepening are changed. The temperature of the gap becomes greater in this case than that in other cases and GB repels with greater force.

As is seen, the processes of colliding GB with the obstacle are complex enough if the exact solution is required. Our purpose was much more modest. We need to show that the optically induced force can take part in processes responsible for bouncing GB on the fax paper.

Having analyzed the most pictorial video [Gooses 2008], we can conclude that there are 2 the most indicative sequences of bounces beginning at time marks I (17, 07), II (21, 23) where the first number in brackets corresponds to the number of the second from begin of the video and the second one corresponds to the number of the frame within the given second (from 0 till 24). It is convenient to designate events in each sequence by time t from beginning the sequence. In this case, the sequences of events are the following

I 0(B), 0.76(T), 1.44(B), 2.00(T), 2.6(B), 3.24(T), 4.00(B), 4.8(T), 5.32(B).

II 0(B), 0.44(T), 0.88(B), 1.28(T), 1.64(B), 2.04(T), 2.56(B), 2.92(T), 3.44(B), 3.76(T), 4.36(B), 4.80(T), 5.16(B), 5.52(T), 6.08(B), where (B) and (T) designate bottom and top GB positions, respectively. The error of each time mark is ±1 frame or ±0.04 s. The average time intervals between adjacent contacts with the surface are t_{JUMP}=1.33 ± 0.02 s and 0.870 ± 0.011s for I, and II sequences, respectively.

On assumption that air resistance is absent, time t of GB falling from height H is expressed as follows

$$t = (2H/g)^{1/2} \qquad (23)$$

If H=0.2 m we have t=0.2 s and time between adjacent jumps is 0.4 s. This is smaller by about 3 times than the time presented in the table. The time can be increased on assumption that the Archimedes force directed upwards compensates partly the force of gravity. In this case, the GB density should be comparable with that of the surrounding air. Stakhanov [Stakhanov 1979] derived the same conclusion at an analysis of bouncing GB on the strip of the conveyer where the time between adjacent jumps is also greater by 3 times than the expected time in accordance with Eq. (23). He assumed that GB consists of gas. However, the well-known motion of the child balloon shows that the resistance of the air slows noticeable the velocity of the ball. Having touched the ground, the ball cannot jump at the same height from which it fell.

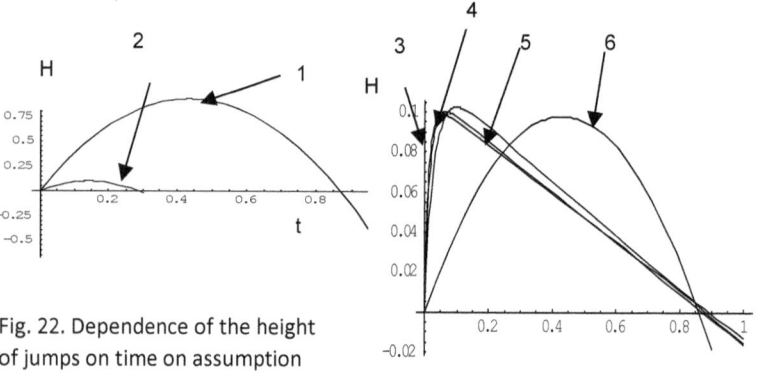

Fig. 22. Dependence of the height of jumps on time on assumption that the air resistance is absent

1) r=0.5 mm, ρ_0=1.0 kg/m³, v_0=1.4m/s, η=0

2) r=0.5 mm, p0=6 kg/m3, v0=0.45 m/s, η=0

Fig. 23. Dependence of the height of jumps on time

3) r=0.25 mm, p0=19 kg/m3, v0=8m/s, η=2 10-5

On assumption that the following three types of forces determine GB motion: gravity force $F_G = 4g\pi r^3\rho_{GB}/3$, Archimedes force $F = 4g\pi r^3\rho_{AIR}/3$, and Stokes forces $F_{STOKES}=6\pi\eta r v$, the GB motion is given by:

$$a = -g(1 - \rho_{AIR}/\rho_{GB}) - \alpha v. \qquad (24)$$

where a is the GB acceleration, g is the acceleration of free falling, r is GB radius, v is GB velocity, $\eta = 20 \cdot 10^{-6}$ Pa s is the air viscosity, $\alpha = 4.5\eta/(\rho_{GB}r^2)$, ρ_{GB} and ρ_{AIR} are densities of GB and air, respectively.

Consider GB motion in time just after it touches the table surface at $t=0$, $h=0$ at initial velocity $v(0)=v_0$. This motion is determined by two parameters α and v_0. Let us pick up these parameters so that the time between adjacent contacts is equal to the time presented in the table.

Numerical solution of (24) for the height h and the velocity v is presented in Fig.22.

This is a single solution. Necessary values $h=0.2$ m and time between adjacent touches $t_0 = 1.1$ s takes place at $v_0=10$ m/s and $\alpha=45$ s^{-1}. One can see in fig.23 that the velocity of GB jumping upwards must be $v_0=10$ m/s, but the velocity of nearing the table surface is $v_1=0.2$ m/s only. Thus, $v_0 = 50\, v_1$. In this case, as is seen in Fig. 23, the time of lifting is significantly greater that the time of dropping. This fact contradicts to all experiments where the times of lifting and dropping are approximately identical.

Thus, we have that, on the one hand, GB is a hollow ball the density of which is comparable with the density of the surrounding air, and on the other hand, GB is not subject to the Stokes force of air resistance. We can propose the following explanation. The Stokes force provides a difference of the pressures produced on the front and rear hemispheres of the GB. In this case, the refractive index near the front hemisphere is greater than that near the rear hemisphere. As a result, the total OIF applied to the GB is directed towards the front hemisphere and compensates completely or particularly the pressure produced by the Stokes force.

Abnormal properties of GBs produced at arc discharges through silicon wafers can be explained on assumption that an intense white light is circulating in a GB shell. The great lifetime of the BL is explained by a perfectly unexpected way. There is the exothermal chemical reaction on the surface of the GB that heats up GB in such a way that the white light arises and is radiated into the GB shell. Such explanation enables to consider GB as a particular case of ball lightning. Both a ball lightning and GB contain white light in a form of whispering gallery waves circulating in a spherical optical lightguide which refractive index is greater than that of surrounding space. Interaction of the light with an inhomogeneous optical medium between GB and obstacle entails a change in the GB velocity. Unlike ball lightning where the spherical optical lightguide is formed by the intense light due to electrostriction compression of air, the spherical optical waveguide in GB is formed by side surface of liquids or solids.

Confinement of the Light at a Strike of Linear Lightning

Having known the BL nature, it is meaningful to consider situations at which BL can appear at strikes of usual linear lightning. The number of electric strikes between clouds is greater approximately by 2 orders of magnitude than number of electric discharges between clouds and the ground. But occurrence BL at electric discharges on the ground has been studied much better.

The basic problem consists in defining those conditions at which light, propagating usually rectilinearly, begin to propagate along closed trajectories inside of a thin spherical layer of compressed air. The analogy with a formation of a soap bubble from a soap film can be useful. As is known, a soap bubble arises if a soap film manages to take an almost closed surface shape.

Let us consider the one-dimensional spatial solitons that are stable films of intensive light propagating in a homogeneous nonlinear optical medium whose refractive index depends on the light intensity {Haus 1984]. If this film is rolled into an infinite cylindrical surface whose axis is perpendicular to the direction of light propagation, light circulates perpendicularly to the generatrices of these surfaces. Then, instead of the infinite cylindrical film, one can consider a film in the form of barrel-shaped surface of revolution with finite length, where light is concentrated in the maximum-diameter section. Finally, similarly to collapse of a hole in a soap film when formatting a soap bubble, two holes at the edges of the barrel-shaped film collapse, and a ball light is formed {Torchigin 2004] Consider how this mechanism operates when BL arises in nature.

A usual lightning stroke is accompanied by the formation of cylindrical sound shock wave due to large current in the lightning core. In this case the air temperature increases up to 25 000-30 000^0 C, and current duration is equal to about 100 μs**Ошибка! Источник ссылки не найден.**. The air pressure in a compression layer (CL) attains 0.3 GPa, the air density is greater than that under normal conditions by an order of magnitude, and the cylindrical wave expands with the velocity of 1,6 10^4 km/s, which is higher than the speed of sound in the atmosphere by a factor of about 50. Under this conditions, the air refractive index in CL is equal $n \cong 1,0027$. It is quite sufficient to form a thin-film cylindrical optical waveguide that can confine completely a light circulating around of an axis of the cylinder of 10 cm diameter that is the typical BL diameter. The light undergoes 1 000 revolutions around the cylinder axis per 1 μs and the cylinder radius increases by 1,6 cm. According to the above estimates, the energy density emitted by a heated air in CL, is insufficient for the formation of an energy clot of hundreds kilojoules.

The processes that can proceed sequentially and/or simultaneously and lead to the accumulation of energy in CL with further BL formation are the following.

1. Accumulation of light energy in CL in time.

2. Concentration of light energy in a domain near maximum diameter of the barrel shaped cylindrical layer (BCL).

3. Self-compression of BCL in width.

4. Collapse of two side holes in BCL that leads to the formation of a spherical layer.

The first process occurs because the CL is optically transparent [Raizer 1992]. The light energy density increase by a factor of 1000 times per micro second because the radiation energy of heated air is added at each revolution to the energy of the light circulating around the ball axis. This energy is accumulated in the CL. In the same time, the energy of the heated air is recovered due to the linear-lightning current.

The second process is connected with features of propagation of light waves of whispering gallery type in BCL (fig.8) whose shape varies in time.

The third process is the self-compression of BCL thickness where the light circulates along closed trajectories. This process is an analogue of known process of self-focusing of the intensive laser beam propagating rectilinearly. In both cases, light propagating in peripheral sections deviated to the side where the refractive index is maximal. In BCL, the refractive index is maximal in the middle of the layer. Self-focusing and self-compression proceed until the thickness of radiation decreases from several centimeters to several micrometers, i.e. by four orders of magnitude. The thickness of layer becomes commensurate with the light wavelengths. Since the volume where the light is propagating decreases, the density of light energy increases accordingly. Joint action of these processes can lead to the necessary increase of the light energy density in BCL.

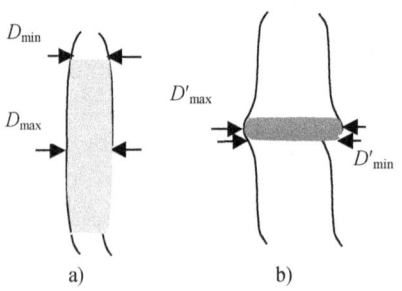

Fig.24. Compression of light radiation in axis direction when the light radiation distributed along essential length (a); concentrated near maximum diameter of the barrel shaped cylindrical layer (b

Let us estimate the self-compression time to be sure that light is not strongly scattered in this time. Let us assume that the areas where the white light is produced by the current of the linear lightning are located in a inhomogeneous medium in which the gradient of the refractive index is directed to the middle of a cylindrical layer and is equal $g_n = 2\Delta n/h$, where $\Delta n \cong 10^{-6}$ is the maximal increase in a refractive index in a middle of the layer, $h \cong 3$ cm is the CL thickness. According to the eikonal equation $d^2 h/dt^2 = 2g_n c^2$. In this case peripheral sections reach the middle in time $t = c^{-1}(h/2g_n)^{1/2} \cong 0.4$ μs. This time is smaller significantly than the lifetime of light radiation in air.

The fourth process that is connected with the collapse of two holes in BCL is not accompanied by concentration of light energy, but is rather important for BL formation. Like the formation of a soap bubble requires a soap film whose hole near the pipe is minimal, the formation of ball lights requires that the maximal and minimal diameters of BCL differ noticeably.

Ought to note that the light intensity near minimal diameter is greater than that neat the maximal one. In this case, the electrostriction pressure neat the minimal diameter is greater than that near the maximal one. The electrostriction pressure tends to near molecules of gas one to another. As a result, the electrostriction pressure tends to decrease the minimal diameter. The same situation takes place in the soap bubble where the surface pressure near the hole tends to decrease the hole diameter.

Interaction of a set of self-confined lights

The cross-section size of fluctuations of gas density that are germs of future light balls are small as compared with diameters of light balls observed by experimenters. The same situation is for AOs. Results of investigation of interaction between AOs and liquid nitrogen show that AO diameter can be smaller than ten micrometers [Klimov 1994]. In the same time experiments on AOs show that there are AOs which sizes are comparable with the cross-section of the set-up used for AOs production. Mechanism of originating of great AOs was considered in [Torchigin 2007]. It was shown that small AOs can merge at their contact and, as a result, AO of greater diameter can appear. Since the physical natures of both AOs and light balls are the same and all these objects can be considered as ball lights, the same effect can take place for light balls too.

Let us show that a steady-state of two identical contacting ball lights (fig.25a) is instable. Suppose that there is some disturbance at which the light intensity in the right bubble is increased. In this case, the air density in its transitive layer increases too and the boundary between the beams propagating to the right and to the left shifts to the left as is shown in fig.25b by dotted line. A part of the light circulating in the left bubble passes to the right one and the initial misbalance increases. As a result, all light of the left bubble passes to the right one. The compressed air in the left bubble remains because transient processes connected with the light are essentially faster than that connected with compression of the air. The first ones are determined by the light speed c whereas the second ones are determined by the sound speed u. As is known, $c/u \approx 10^6$. Having lost the electrostriction pressure exerted by the light, the compressed air in the left bubble begins to expand. As a result, a crash can be heard. As for the right bubble, that the increase in the light intensity entails an increase in the air pressure in the transitive layer by means of attraction in the bubble shell additional air molecules from surrounding space. As a result, the bubble diameter increases.

It is easy to check, that two contacting bubbles of different diameters with identical light intensity are instable too. The light passes to the bubble with the greatest diameter. The same is valid

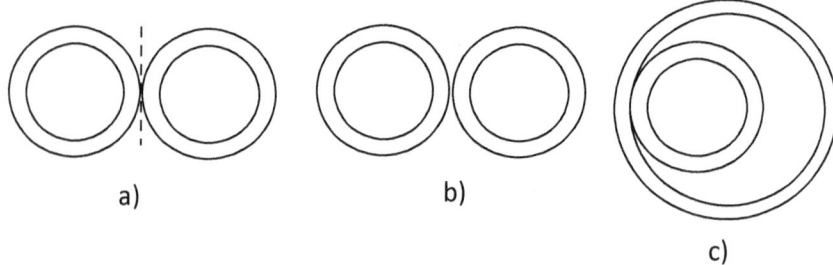

a)　　　　　　　　　　　b)

c)

Fig.25. Instability of two identical ball lights (dotted lines show boundaries between light beams propagating in different bubbles. (a) two identical bubbles; (b) light intensity in the right bubble is greater than that in the left one; (c) light intensities are similar in the bubbles

for bubbles shown in fig.25c. Thus, a relatively great final ball light that is essentially greater than initial bubbles can be formed in a process of gas discharge.

There are the following evidences in favor of merging small ball lights. These bubbles are characterized various radiation losses for various wavelengths. It is connected with the fact that a ratio d/λ is relatively small for them. Here d is the bubble diameter, λ is the wavelength of the light circulating in the bubble shell. Radiation losses in such bubbles increase with a decrease in the ratio d/λ**Ошибка! Источник ссылки не найден.** [Oraevskiy

2002] As a result, light waves from the red region of the spectrum leave the bubble shell faster than light waves from the blue region and bubble color shifts in blue side of the spectrum. That is why liquid nitrogen with small ball lights penetrated within it is blue-green in color. Since a source of light waves circulating in a shell of the great final bubble is waves of small ball lights, a color of the final bubble should be moved in the blue side of spectrum too. Indeed, the shell of great bubble in experiments with ultrasonic streams is violet in color [Klimov 1993].

In all experiments on production of ball lights with small lifetime beginning from Plante experiments, presence of a crash that accompanies chaotic movement of arising luminous spheres is marked. Such crash can be caused by expansion of an air compressed in AO that has merged with other bubble. The compressed air in the shell remained after merge of a bubble begins to extend sharply, causing a crash. Something similar crash is heard at electric welding. It is not excluded, that the crash arising at it has the same nature.

Mechanism of appearance of ectons and cathode spots

There are strong reasons to believe, that features of vacuum discharges can be explained by an existence of Ball Light. The vacuum in the discharge gap takes place only at the initial stage of the discharge. At a steady state vapors of the metals which have evaporated from the cathode reside in the discharge gap. As is known, the vacuum discharge can be accompanied by occurrence of avalanches of electrons, each of which contains 10^9-10^{11} electrons. Similar avalanche has been called by ecton [Mesyats 2000]. The occurrence of such avalanches can be explained as follows. Intense light in plasma from metal vapors is subjected to self-organizing. This process results in the appearance of Ball Lights shell of which consists of metal vapors [Torchigin 2004 Doclady Physics]. Interaction of AO with metal surfaces has been investigated experimentally [Avramenko 1994]. It has been shown, that the AOs are attracted to such surfaces. They can either burn a hole in such surface, or burn out a crater if the energy of AO has not sufficient.

This phenomenon can be easily explained [Torchigin 2004 Opt Comm.]. Having come nearer to the surface of the cathode, the Ball Light evaporates metal as the Ball Light heats up the surface owing to the light radiated by Ball Light. Having located in the area with the maximal density of gas, the Ball Light evaporates the metal until its energy exhausted. The evaporation of the metal and heating of the cathode are accompanied by the usual phenomenon of the thermo emission, used in radio tubes. When the energy of Ball Light becomes smaller than a threshold, Ball Light disappears. As a result, being near the surface of the cathode, the Ball Light causes the issue of an avalanche of electrons. We should notice that the character of the craters formed by Ball Light on metal surfaces at erosive gas discharges [Avramenko 1994] and the craters formed on the surface of the cathode at vacuum discharges [Mesyats 1995] are identical. The description of the casual movement of the fiery sphere in the Plante experiments reminds the description of the casual movement of the cathode spot at vacuum discharges [Luybimov 1978]. Probably, the Ball Light is responsible not only for an occurrence of ectons, but also for an occurrence of cathode spots.

Self-confinement of light in plasma

As is known, the permittivity ε of plasma depends on the concentration n_e of electrons within it and is determined as follows

$$\varepsilon = 1 - \frac{4\pi e^2 n_e}{m\omega^2}$$

(25)

where e and m are the electron charge and mass, respectively, $\omega = 2\pi/\lambda$, λ is the wavelength of light for which ε is determined. As is seen, ε depends on the electron concentration n_e and does not depend on the ion concentration n_i. Ought to distinguish plasma in time of a gas discharge and in time when the gas discharge is completed. In the first case, there is a steady-state process where a recombination of plasma is compensated by an ionization of air atoms due to the gas discharge. In the second case, plasma disappears gradually. As was shown, the lifetime of plasma is about several milliseconds only [Raizer 1980]. That is why all theories where plasma is used for explanation of the BL phenomenon are anxious about plasma lifetime. In our opinion, this anxiety is unjustified because there is usually no plasma within Ball Light after its generation. There is certain no plasma after penetrating Ball Light through windowpanes because plasma cannot penetrate through glass. On the contrary, there are no doubts that plasma exists in time of Ball Light originating and it may play essential role in the process of Ball Light generation.

Under assumption that all atoms of a gas are ionized and a number of free electrons is equal to a number of atoms in the gas and equal to $n_e=5 \ 10^{19}$ cm^{-1} we obtain from (25) that at normal conditions and $\lambda=0.6$ μm the refractive index of such completely ionized plasma is smaller than 1 by 1.6%. For example, we can imagine Ball Light shell with $n=1$ merged in the plasma with $n=0.984$. A change in the refractive index is greater by about 50 times than that obtained at a separation of gas mixture.

As was shown above, a light propagating in a gas mixture tends to increase the refractive index in the region where it propagates. As applied to plasma, the light tends to decrease n_e within Ball Light shell, that is, to push out electrons outside. In this case, a homogeneity and neutrality of plasma violate. As for neutrality that positive charged ions are attracted to electrons and are pushed out too. They are substituted by neutral atoms. As a result, the refractive index n in the layer located outside Ball Light decreases. Dependencies of the refractive index n as well as concentrations n_e and n_i on the distance r from the Ball Light center are shown in fig. 26. A spherical layer where an intense light propagates is dotted.

The energy required for such separation of plasma is minimal because plasma remains neutral. It is not required to overcome forces of attraction or repulsion between electrons or ions. This

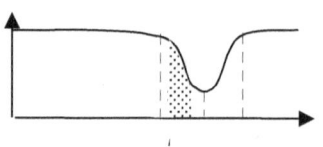

Figure 26. Dependence of plasma refractive index in the BL shell on the distance from its center:

situation reminds separation of molecules of gas mixture where molecules of the component with minimal n are pushed out Ball Light. The region $R_0-\Delta R<r<R_0+\Delta R$ in fig. 26 is a tunnel [Kogelnik 1988] which confines the light circulating within Ball Light and prevents it from radiation in free space. The light is concentrated in the region $R_0-\Delta R_1-w<r<R_0-\Delta R_1$ where $\Delta R_1>w$, w is the Ball Light thickness.

Ought to underline that Ball Light exists in time when the plasma exists. The plasma exists when the gas discharge takes place. This situation corresponds to the experiments in begin of the 20th century by Plante and Gezehus [Gezehus 1900].

Confinement of electrons by circulating light

The following situation may be imagined. A thin spherical layer of compressed air provides confinement of an intense light circulating within it. The light can provide confinement of a layer of electrons located within the spherical layer of compressed air from expansion because the light prevents a decrease of the refractive index in the region where it is circulating. Since electrons within Ball Light shell decrease its n, they are pushed out from it. Thus, Ball Light can confine negative charges from dispersion due to mutual repulsion, and, therefore, negative charged Ball Lights may exist. Such Ball Lights, like any charged things, can be attracted to metal objects until other factors come into force.

Electrons located with Ball Light shell tend to expand and produce the pressure P_E at the inner surface of the Ball Light shell. The pressure P_E decreases the electrostriction compression exerted by the light circulating in Ball Light shell.

Let determine a relation between P_E and charge Q in the Ball Light shell. The energy E of spherical capacitor of R radius and the pressure P_E exerted by charges in the Ball Light shell are equal, respectively [Nikitin 2002]:

$$E = \frac{Q^2}{8\pi\varepsilon_0 R},$$

$$(26)$$

$$P_E = \frac{E}{SR}$$

$$(27)$$

where S is side surface of the shell. Pressure P_E tends to expand the shell of volume $V=S\Delta R$ because discharges are repelled one from others. Here ΔR is the thickness of the shell. On the other hand, the energy density of compressed gas is expressed as $W_G=P_G/\gamma$. In this case energy of compressed gas is $E_G=W_G V= S\Delta R\, P_G/\gamma$ and

$$\frac{P_G}{P_E} = \frac{R}{\Delta R}\frac{\gamma E_G}{E}$$

$$(28)$$

If $E=E_G$, the pressure exerted by compressed gas is greater by a factor of $\gamma R/\Delta R$ than the pressure exerted by negative charges. It is not surprising because we assumed that the gas is concentrated in a thin layer of ΔR thickness. Let determine the pressure $P^\perp$ that the gas exerts at a shell of R radius. At increasing R by δR the mechanical work produced by the pressure $P^\perp$ is equal to $P^\perp S\delta R = P^\perp 4\pi R^2\delta R$. In the same time the mechanical work produced by the compressed gas is equal to $PdV/dR\delta R$, where $V=4\pi R^2\Delta R$. From equality of these works we have $P^\perp/P=2\Delta R/R$ and from (28)

$$P^\perp/P_E=2\gamma$$

$$(29)$$

This means that the energy of confined charges can be comparable with that of compressed gas. If for any reasons a hole in Ball Light shell appears in a form of area where no intensive light is propagating, compressed electrons are pulled out the Ball Light shell in free space immediately. This accident can be accompanied by explosion and strong electromagnetic indignation.

Accelerating decay for production of energy

One can found by studying various publications in Internet, accessible by keyword «ball lightning» that the term «Low Energy Nuclear Reactions" (LENR) arises in the same publications where the term "ball lightning" is used. Moreover the International conferences devoted to the ball lightning is called "Ball Lightning and «Low Energy Nuclear Reactions". Undoubtedly, the well-known phenomenon of the natural radioactivity is LENR where a nucleus of the radioactive element ceases abruptly its existence. This phenomenon is accompanied by a release of the energy. This energy can be used in practical applications. For example Plutonium-238 has a half-life of 87.7 years, reasonable power density of 0.54 watts per gram and is used as a power supplier for satellites.

However, natural radioactive materials with relatively small half-life are absent on the Earth because they are long gone. We can imagine attempts to decrease the life time of elements, life time of which is great enough and which exist on the Earth. One of these attempts was successful. Nuclear plants produce a noticeable part of the energy for mankind due to the chain nuclear reaction. However, these plants are dangerous and should be cumbersome.

On the other side, successful attempts to decrease the life time by means of great pressure, temperature, powerful electric discharges are known. These attempts are accompanied by arising small luminous objects that are considered as miniature ball lightnings [Levis]

A natural assumption arises that physical conditions within Ball Lightning are favorable for decreasing life-time. However, at present, a generally accepted notion about physical conditions within Ball Lightning is absent. Therefore, it is not clear that physical conditions are required.

Physical Conditions in Experiments where Nuclear Conversions are observed

It has appeared that quickly prepared long living gas of densely packed molecules took place in all experiments. Ought to note that the shell of the ball light consisting of strongly compressed air can appear extremely quickly during a fraction of a microsecond. There are some experimental confirmations of this fact [Urutskoev 2002]. It is no wonder because the light participates in its formation and makes thousand revolutions in the shell in this time. Thus, physical conditions inside of an AO shell are suitable for increasing the speed of the radioactive decay. AO shell is, probably, a single object among known ones which provides necessary physical conditions at which time of interaction $t > \tau_1$. Analysis of experimental studies where a mention about nuclear conversions takes place shows that AOs are observed in the great bulk of these studies. There is one group of the studies where nuclear conversions are observed but direct mentions about AOs absent. But a quickly prepared long living gas of densely packed molecules is presented in these studies too. Consider results of some experiments.

At first sight, no such gas existed in classical Pons and Fleishman experiments connected with electrolysis of palladium electrodes in heavy water. Really, it seemed so within 14 years. In 2003, as a result 14-years research the following facts [Storm 2003] have been established. First, areas in which nuclear activity takes place lay on a surface of the cathode, instead of inside of metal as it was supposed earlier. Secondly, these areas exist in dendrites or nano crystals, located on a surface of an electrode, instead of on the very surface. A conclusion has been drawn that the only thing necessary for a nuclear conversion is existence of the small isolated domains of a material. It can be dendrites growing on a surface of the electrode or impregnations in the palladium electrode, or grains of other material on the palladium electrode. Erosive discharges on these inhomogeneities can occur

These conclusions were confirmed in Szpak work [Szpak 2003], where was shown that nuclear conversion is accompanied by occurrence on a surface of electrode of "hot spots" and micro explosions. It was estimated that each micro explosion is accompanied by a nuclear conversion in which from 10^4 up to 10^9 nuclei participate. It is not excluded, that the reason of micro-explosion is disappearance of a miniature ball lightnings have appeared as a result of local erosive discharges. The term "miniature ball lightning" means the same that our term "autonomous object". We shall use already accepted term though the term miniature ball lightning is more often used. Japanese investigator Matsumoto [Matsumoto 1995-2001] repeatedly informed about observation of AOs in conditions of classical Pons and Fleishman experiments

Lewis analyzed surfaces of those cathodes that have been used in successful attempts to carry out nuclear conversion. It turns out that these cathodes, unlike other cathodes, have been undergone to strong erosion [Lewis 2003]. Besides, occurrence of AOs from their surface was observed in a process of nuclear conversion. Similar erosion of electrodes is observed in numerous experiments on AO production by means of erosive gas discharge [Avramenko 1994]. Thus, at present time it is authentically established that nuclear conversion in conditions of classical experiment is accompanied by occurrence AOs and, therefore, there is a quickly prepared long living gas of densely packed molecules.

Such gas presents also in cavitation bubbles produced in liquids by means of an intensive acoustic wave. It is known, that pressure in such bubbles at their compression can reach several thousand atmospheres. As was shown in Kladov experiences carried out in 1998-2001, nuclear conversion took place in such bubbles [Kladov 1998-2001]. Analogous experiences have been carried out by Taleyarkhan group in 2004. Now this group is financed by DARPA.

The conclusion that the gas obtained at explosions accompanied by erosion of a material is favorable for nuclear conversion has been confirmed by Ukrainian scientists who used a powerful pulse of electrons with energy about 1 kJ for super compression of substance [Adamenko 2004].

It would seem, the gas is absent in Correa experiments connected with production of excess energy by means of vacuum discharges. [Carrel 1996]. However, works of other researchers, in particular, Shoulders showed that the vacuum discharge is accompanied by occurrence of so-called exotic vacuum objects and transmutation of elements [Shoulders 2005]. In reality, such objects represent AOs, and their shape where intensive light circulates consists of vapors of metal that are evaporated from the cathode of a vacuum tube. Thus, the gas consists of molecules of metal in this case. The same is valid for Solin experiments with welding of zirconium in vacuum by electronic beam [Solin 2001].

By the way, Sholders studies allow to explain existence of ectons which are avalanches of electrons arising at the vacuum discharge [Mesyats 2000]. An appearance of an avalanche can be explained somehow within a frame of usual representations without attraction of AO but it is impossible to find out reasonable reasons caused a disappearance of an avalanche. As was shown in chapter 2, introduction of AOs into consideration of the phenomena at vacuum discharges enables to explain both occurrence, and disappearance of avalanches, and also the fact of their existence. Introduction of AOs into consideration permits also to explain occurrence of excess energy in Correa experiences because any appearance of AOs is accompanied by appearance of a quickly prepared long living gas of densely packed molecules. The excess energy appears due to acceleration of nuclear decay. Indeed, transmutation of elements has been fixed in Sholders experiments.

R.F. Avramenko has informed about generation of excess energy by means of AOs in a form of so-called great energy plasma formations [Avramenko 1994. In this case, AOs were produced by erosive gas discharge in air at normal atmosphere pressure. Urutskoev has informed about a transmutation of elements at explosions of titanic wires at discharge of the battery of capacitors about 50 kJ stored energy [Urutskoev 2002]. He has marked that the explosions are accompanied by occurrence of "spherical plasma formations". In reality, they are AOs. By the way, the fact of extremely fast occurrence of ball lights has been fixed in these experiments in the first time. It has appeared, that at shooting by high-speed video camera a ball light appears so quickly, that it is absent on one frame, and it has been generated completely on the following frame.

Thus, introduction of AOs into consideration allows finding out some common and explaining seemingly absolutely different experiments. All known experiments on successful realization of nuclear conversions, where either generation of excess energy, or transmutation of elements takes place, are accompanied by an appearance of a quickly prepared long living gas of densely packed molecules.

Some views on order of magnitude constants used above can be obtained from analysis of physical conditions of successful experiments accompanied by nuclear conversions. Since life time of AO is in the interval from 1mks to 1 ms and the lifetime is comparable with the time constant τ_{NC} for passing a system of densely packed oscillators in a steady state, then τ_{NC} is in an interval from 10^{-6} up to 10^{-3} s. The beating period is greater by 1-2 orders of magnitude than τ_{NC} and, hence, and τ_b is in an interval from 10^{-5} up to 10^{-1}s.

According to modern representations of nuclear physics "in a nucleus, considered in a form of a drop [Pik—Pichak 1990], there are oscillations with the period $\tau_n = 10^{-21}$s and amplitude equaled to 0,1-0,2 radius of a nucleus ". Taking into account that $\tau_n \approx 10^{-21}$ s, we obtain from (13) $k = \tau_n/\tau_b \approx 10^{-20}$–$10^{-16}$. Certainly, these evaluations are only an illustration of how extremely small coupling index can lead to an appreciable result for reasonable time.

Observation of appearing AOs testifies that physical conditions in AOs are favorable for acceleration of nuclear decay. Since the nature of AOs is known, these conditions can be created in laboratory. It is known, that experiments with nuclear conversions are characterized weak reproducibility. Low reproducibility takes place also in experiments on AOs production because a physical nature of AOs was not known for experimenters. Purposeful use of erosive gas discharge allows to eliminate the specified lack and to start purposeful accomplishment of nuclear conversions. In accordance with one of hypotheses, the Ball Lightning was a reason of Chernobyl tragedy [Torchigin 2006], which essentially delayed development of atomic engineering. It is not excluded, that the Ball Lightning will expiate the fault and will play solving positive role in development of absolutely new direction in nuclear energetic which will lead to development of new cheap alternative energy sources. How the life on the earth will be changed is described by American futurologist Arthur Clark [Clarke 1992].

Possible Types of Nuclear Decays

From the presented consideration follows that interaction of only identical nuclei can accelerate a nuclear decay. Besides, the smaller the difference between the energy stored in a nucleus before decay and after decay the easier conditions for realization of such conversion can be fulfilled. What types of nuclear conversions can be carried out by means of accumulation of interaction between adjacent identical nuclei? It is not difficult to find out a majority of possible decays in which the minimal excess energy takes place. In this case the total number of protons and neutrons is kept after conversion.

Similarly to classical reaction of synthesis of deuterium nuclei

$$2\,{}^{2}_{1}D \rightarrow {}^{4}_{2}He + 23{,}8 \text{ MeV},$$

We can write, for example, the following reactions

$$2\,{}^{27}_{13}Al \rightarrow {}^{26}_{12}Mg + {}^{28}_{14}Si + 3{,}315 \text{ MeV},$$

$$2\,{}^{63}_{29}Cu \rightarrow {}^{64}_{30}Zn + {}^{62}_{28}Ni + 1{,}59 \text{ MeV}.$$

The excess energy is determined by the difference in mass defects ΔM between initial and resulting products. In accordance with [Grigoriev 1991] ΔM for considered nuclei are the following:

ΔM for ${}^{27}_{13}Al$ $-17{,}194$ MeV,

ΔM for ${}^{26}_{12}Mg$ $-16{,}212$ MeV,

ΔM for $^{28}_{14}Si$ – 21,491 MeV,

ΔM for $^{63}_{29}Cu$ – 65,578 MeV,

ΔM for $^{64}_{30}Zn$ – 66,001 MeV,

ΔM for $^{62}_{28}Ni$ – 66,745 MeV.

Notice, that the left hand side in the record of reactions begins with number 2. It means that identical nuclei can participate in reaction only. The first reaction with deuterium is reaction of synthesis. However, it is impossible to tell the same about the second and third reactions as two nuclei exist before and after reaction. After the analysis ΔM for some heavy nuclei, it is possible to write, for example, the following reaction:

$$2\,^{206}_{82}Pb \rightarrow \,^{206}_{82}Pb + 2\,^{103}_{41}Nb + 102.23\,MeV$$

This is typical reaction of radioactive decay with generation about 100 MeV additional energy. Actually there is a splitting Pb nucleus on 2 identical Nb nuclei by means of other Pb nucleus that is accompanied by generation of superfluous energy. Certainly, splitting can occur and on not identical nuclei. Thus, there can be a synthesis of nuclei, their splitting, and also transmutation of elements, at which two identical nuclei are transformed in various nuclei. Thousands of similar reactions can be imaged. The considered examples are only an illustration. The analysis of possible types of nuclear conversion is a problem of nuclear physics. We would like to note only one obvious conclusion. The smaller life time of a nucleus or excess energy are the easier to carry out nuclear conversion with such nucleus. Indeed, in this case there are either relatively great radiating losses or/and a small difference E_0-E_{th}.

An isotopic conversion is accompanied by small excess energy. What is why a change of a natural isotopic ratio is mentioned often. It is very pictorial experiments of Urutskoev at explosion of a titanic wire that were accompanied by appearance of AOs. Natural isotopic ratio changed after the explosion as follows. A fraction of isotope Ti[48] decreased from 72 % up to 62 %, percent of isotopes Ti[46], Ti[47], Ti[49], Ti[50] increased from 8 % up to 10 %, from 6 % up to 8 %, from 10 % up to 12 %, from 6 % up to 8 %, respectively. Unfortunately, the paper with these data disappeared from Internet. Mention about these reactions is in [Vysotskii 2003]. Apparently, there were nuclear conversions of the following type

$$2_{22}^{48}Ti \rightarrow _{22}^{49}Ti + _{22}^{47}Ti + 3,485 \ M\mathfrak{s}B$$

or

$$2_{22}^{48}Ti \rightarrow _{22}^{50}Ti + _{22}^{46}Ti + 0,420 \ M\mathfrak{s}B$$

In the first case one neutron passes from one nucleus in others and in the second case two neutrons pass.

Probably, similar reactions are used by colonies microbiological cultures in radioactive waste with rather small life time. There is evidence that in such environments transmutation of elements is possible. Besides, the speed of radioactive decay noticeably increases [Urutskoev 2002]. Thus there are neither explosions, nor intensive light, nor any AOs.

"Micro installation" in a form of a bacterium can be imagined which have learned to break a stationary phase relation between adjacent oscillators by sharp relocation of oscillators. As a result, the coupling index between oscillators and their own frequencies change. It causes phase shift of oscillations in these oscillators relative a phase of oscillations in an isolated oscillator. When the phase shift reaches $\pi/2$, the bacterium nears the oscillator with other oscillator having the steady state phase. As a result, the phase shift between these oscillators becomes equal to $\pi/2$. This provides transmission of energy from one oscillator to others. Thus the bacterium accelerates a decrease in energy of nucleus until its energy achieves threshold E_{th} and the nucleus breaks up. Excess energy generated at this process is used by bacterium for ability to live.

As a rule, coupling index between usual nuclei is so small and the difference E_0-E_{th} is so great that the bacterium is not able to provide a necessary decrease in E_0 down to E_{th}. But the situation can be favorable for nuclei with rather small half-life period. There is double usefulness from small half-life period of a nucleus. Firstly, the smaller half-life period of a nucleus the greater the amplitude of wave function outside the nuclei because in accordance with generally accepted notions the square of the wave function outside a nucleus is proportional to the probability of its decay. In this case coupling index between adjacent nuclei is greater than that between nuclei with great half-life period. Secondly, the difference E_0-E_{th} is small enough because the excess energy is small. Radioactive waste with rather small life time contains such nuclei. Bacteria have learned to extract and use excess energy for their life earlier, than the people have. Bacteria have advantage over people because there are no theorists among bacteria who assert that it is impossible.

Certainly, it is very hard work to find out the most suitable new fuel for new alternative sources of energy. On the one hand, it should be a wide spread element with relatively long life time so that it is not dangerous for health. On the other hand, its excess energy at accelerated nuclear decay should be considerable and conditions for production of a quickly prepared long living gas of densely packed molecules should be as simple as possible. Thus, instead of a single nuclear reaction of synthesis of deuterium nuclei which takes place in H-bomb there are a lot of other nuclear reaction which are also accompanied by appearance of excess energy. Possibly, presented above our rough consideration can determine a direction of further investigations. Ought to underline that there is no mystics. Indeed, natural radioactive decay is an objective reality. Possibility of changing the speed of radioactive decay is also objective reality. Appearance of excess energy and transmutation of elements are also objective realities. Everything that is required it is to find out a way of using these realities in practical applications.

On assumption that physical conditions within Ball Light are favorable for production of nuclear reaction, we can conclude that the following requirement should be fulfil. It is necessary to pack molecules of gas as quickly and densely as possible to provide condition that the relative positioning of the molecules would be kept in a relatively long time.

Certainly, the fast decay can be used as a new alternative way for production of energy.

Conclusion

We have presented a set of new physical phenomena that we have managed to disclose without leaving a computer. We took into account the thought of the Russian Nobel prize winner Peter Kapitsa that Ball Lighting is a small window in the great world of new physical phenomena. Having observed through suppositive window in a form of Ball Lightning, we clap eyes on a lot of unknown physical phenomena that are independent on the Ball Lightning. The most significant among them is an existence of the self-confined light in the conventional air atmosphere. We can not forecast when this phenomenon will become generally recognized. At least, the progress over ten years since time when this phenomenon was put forward shows that the scientific community is not ready to the recognition. This fact can be explained easily. A common sense protests against the thought that the conventional white light can confine itself in the conventional air atmosphere. All these objects are familiar very well to everyone and his experience tells that the light travels in straight lines.

Similar situation took place with the Copernicus theory. Everyone can see that the Sun moves around the Earth rather than vice versa. About century was required to recognize the Copernicus theory. We hope that smaller time will be required to recognize an existence of the self-confined light.

References

Abrahamson, J.,Dinnis, J. Nature2000, 403, 3, February,519-521.

Adamenko S. V., Adamenko A. S., Vysotskii V. I., 2004. Infinite Energy, 9 (54), 23–30.

Avramenko R. F, Nikolaeve V. I., Poskacheva L. P In book Ball lightning in laboratory. Avramenko R. F. Ed.; Himiya, Moscow, 1994, pp.7-56 (in Russian).

Avramenko R. F.Ed. Ball Lightning in a laboratory, (Moscow, Himiya, 1994).

Barry J. D., Ball Lightning and Bead Lightning, Plenum Press: NY, 1980.

Carrel, Mike, (1996). In Infinite Energy Magazine Special Selection pp. 62–70.

Clarke A. C., (1992). The coming age of Hydrogen Power. Infinite Energy Magazine Special Selection, pp.8–10.

Emelin S. E. et al (1994) Physical conditions of the ball lightning ejection caused by interaction of electrical discharge with metal and polymer; http//www.balllightning.narod.ru/isbl01/BL_eject.htm

Emelin S. E. et al J. Tech Fiz, 1997, 67, № 3, 19–28.

Gezehus N. A. Journal of Russian chemical physical society, 1900, 8, 311.

Gooses V., P. de Graaf and R. Dekker
http://www.youtube.com/watch?v=QLTPELhKAYM and
http://www.youtube.com/watch?v=QLTPELhKAYM&mode=related
&search

Haus H.A. Waves and Fields in Optoelectronics (Prentice Hall, New Jersey) 1984.

Kladov A. In 13-th Radiochemical Conference. 19–24 April, 1998. Marianske Lazne Jachymov Czech Republic. Booklet of Abstracts.

Kladov A. In 21-th International Symposium «Industrial Texicology 2001». Proceedings. 30 May – 1 June 2001, Bratislava, Slovak Republic.

Kladov, A. In 5-th International Conference on Nuclear and Radiochemistry. Pontresina, Switzerland, 3–8 September, 2000, Extended Abstracts vol. 1.

Klimov, A. I.; Mishin, G. I. Letters in J. Tech. Fiz. 1993. 18 (13), 19.

Klimov, A., I.; Malchenko, D. M.; Sukovatkin, N.,N. In Ball lightning in laboratory; Avramenko R. F.; Ed.; Himiya: Moscow, 1994.

Kogelnik H. Theory of optical waveguides in book Guided wave optoelectronics edited by T. Tamir (Springer-Verlag, Berlin, 1988)

Landsberg, G. S. Optics Nauka: Moscow, 1976.

Lewis, E. H., In Tenth International Conference on Cold Fusion, USA, Massachusetts, Cambridge, August, 2003.

Luybitov, Yu. N. In Physical encyclopedia, Prokhorov, A., M.; Ed.; Bolshaya Rosiyskaya Ensiklopediya: Moscow, 1998, Vol. 1, pp 375–379.

Matsumoto T., Fifth International Conference on Cold Fusion. 1995 April 9–13, Monte Carlo, Monako.

Matsumoto, T. Fusion Technology, 1992, 22, 281.

Matsumoto, T., 2001. IEEE International Pulsed Power Conference, 2001, 1, 273–276.

Mesyats, G. A. Ectons in vacuum discharge: discharge, spark, arc; Nauka: Moscow, 2000.

Mesyats, G.A., 1995. Uspehi Physics;1995 165 (6), 601–626.

Naschokin, V. V. Technical thermodynamics and heat transmission; Visshaya shkola: Moscow, 1969.

Oraevskiy, A. N. Quantum electronics; 2002, 32, № 5, 377–400.

Pik-Pichak, G., A., In Physical encyclopedia, Prokhorov, A., M.; Ed.; Bolshaya Rosiyskaya Entsiklopediya: Moscow, 1990, Vol. 2, pp 238–239.

Rizer Yu.P. In Physical Encyclopedia; Prokhorov A.M.; Ed.; Bolshaya Rossiyiskaya Ensiclopedia: Moscow, 1992; Vo3. 1, pp 448.

Rizer, Yu. P. Foundations of modern physics of gas discharges; Nauka: Moscow, 1980.

Shoulders, K. Infinite Energy, 2005, 61.

Solin, M. I. Physical Thought in Russia; 2001, no.1, 43–58.

Spillane, S. M.; Kippenberg, T. J.; Vahala, K. J. Nature; 2002, 415, 621–623.

Stakhanov I. P. The physical nature of Ball Lightning (Atomizdat, Moscow 1979 CEGB trans CE 8244)

Storms, E. Tenth International Conference on Cold Fusion, USA, Massachusetts, Cambridge, August 2003.

Szpak, S;, Mosier-Boss, P. A., Dea, J., Gordon, F. Tenth International Conference on Cold Fusion, USA, Massachusetts, Cambridge, August, 2003.

Taleyarkhan, R. P. et al Journal of Power and Energy; 2004 218 (5), 345–364.

Torchigin V. P. On the nature of Ball Lightning, Doclady Physics vol. 48, no. 3 pp. 108-11 (2003).

Torchigin V. P., A. V. Torchigin, 2005 Physical Nature of Ball lightning. European physical Journal D 36, (2005), 319–327.

Torchigin V.P., A.V. Torchigin, Features of Ball Lightning stability, Europhysics Journal D 2005, 32, 383–389.

Torchigin V.P., A.V. Torchigin, Phenomenon of ball Lightning and its outgrowth. Phys. Lett. A; 2005, 337, 112–120.

Torchigin V.P., Torchigin A.V. Behavior of self-confined layer of light radiation in the air atmosphere. Phys. Lett. A. 2004, 328/2–3, 189–195.

Torchigin, V. P. Manifestation of Optical Quadratic Nonlinerity in Gas Mixtures. Physics; 2004, 49, No.10, 553–555

Torchigin, V. P., Torchigin A. V., Space soliton in gas mixtures. Opt. Comm. 2004 240/4-6, 449-455

Torchigin, V. P., Torchigin, A. V. Mechanism of the Appearance of Ball Lightning from Usual Lightning. Doclady Physics; 2004, 49, No. 9, 494–495

Torchigin, V. P., Torchigin, A. V. Propagation of self-confined Light radiation in Inhomogeneous Air. PhysicaScripta, 2003, 68, 388–393.

Torchigin, V. P., Torchigin, A. V. Self-organization of intense light within erosive gas discharge. Phys. Lett. A; 2007, 361, 167–172.

Vysotskii, V. I.; Shevelev, V.N.; Tashirev, A. B., Kornilova, A. A., Tenth International Conference on Cold Fusion, USA, Massachusetts, Cambridge, August, 2003.

Urutskoev, L. I. Lomonosov; 2002, 10, 8-12 (In Russian).

Urutskoev, L. I., Fillipov D. N.. Usp. Fiz. Nauk; 2004, 174, № 12, 1355–1358.

www.ingramcontent.com/pod-product-compliance
Lightning Source LLC
Chambersburg PA
CBHW070232210526
45168CB00020B/2037